The King's

MATTER
MIND AND MEANING

by

WHATELY CARINGTON

With a Preface
by
H. H. PRICE

Essay Index Reprint Series

BOOKS FOR LIBRARIES PRESS
FREEPORT, NEW YORK

First Published in the United States in 1949 by
Yale University Press

Reprinted 1970 by arrangement with Methuen and Co., Ltd.

STANDARD BOOK NUMBER:
8369-1596-8

LIBRARY OF CONGRESS CATALOG CARD NUMBER:
78-111818

PRINTED IN THE UNITED STATES OF AMERICA

CONTENTS

PREFACE

The death of Mr. W. Whately Carington was a most grievous loss to Psychical Research; not least because he did not live to finish this book, to which I know he attached great importance. Fortunately he had written the first five chapters, more or less in their final form. I have only found it necessary to add a few footnotes and cross references. But with the last chapter, 'Mind and Matter', which, as he tells us himself, was to be 'in a sense the core of this book',[1] the position is much less satisfactory. There were only a few fragmentary pencil notes, written during his last illness. Apparently there were two alternative versions, both very brief, which I have conflated as best I could; and Ch. VI, as now printed, is the result.[2] I have thought it well to add three brief essays as appendices. I have chosen them from among a number of papers which Mrs. Carington kindly placed at my disposal, because they seem to me to illustrate some of the main themes of this book. Appendix I, 'Don't shoot the philosophers—yet', is a more popular version of Ch. II, 'The Failure of Metaphysics'. Appendix II, 'Life after death, the need for an inversion of thought', should be compared with Ch. IV, 'Mind'. (Cf. also Carington's earlier book, *Telepathy*, Ch. XI.)[3] Appendix III, 'Does to-morrow exist?' is a very ingenious and somewhat tantalizing sketch of a theory of Precognition, the most puzzling, perhaps, of all supernormal phenomena. I have included it partly for its own intrinsic interest, and partly because it illustrates the theory of normal perception which is stated in Ch. IV.

The book may be described as a contribution to the philosophy of Psychical Research. The development of Psychical Research has hitherto been somewhat one-sided. On the factual side great progress has been made in the last sixty

[1] p. 228; last words of Ch. V.

[2] I shall make some tentative suggestions later about the missing parts of Ch. VI. See pp. xvi–xx, below.

[3] Methuen, London, second edition. 1945.

years. A very large mass of evidence has been collected (how large, a glance at the published proceedings of the S.P.R. and similar bodies will show). Much of it is of very high quality, and some of it has been obtained under laboratory conditions.[1] But very little progress has been made on the theoretical side. We have a mass of queer facts, but as yet we have no satisfactory conceptual framework to correlate them with each other and render them intelligible; still less any conceptual framework which will bring the 'supernormal' facts into some intelligible connexion with the 'normal' ones established by the orthodox sciences. Indeed, the very use of such words as 'supernormal' and 'paranormal' is a kind of confession of this. We call an occurrence 'supernormal' because it is something which *ought* not to happen if our ordinary outlook on the world is correct. We need to revise our fundamental concepts—and especially our concepts of mind, of matter, and of the relations between them—in such a way that these queer facts will not seem queer any longer, but on the contrary will appear to be what might reasonably have been expected. Incidentally, the facts themselves will not be generally accepted until we do. Events which apparently 'make no sense' and 'don't fit in anywhere' naturally tend to be ignored for that very reason. However strong the evidence is, they make no permanent impression on the mind, and particularly on the educated mind; and even if their reality is admitted, they tend to be quickly forgotten.

Such a revision of fundamental concepts is the object of this book. The author's aim, as he tells us at the beginning of Ch. I, is 'to clean up once for all—in principle and in outline at least—this great muddle about the relation of Mind to Matter, which has fretted philosophers ever since philosophizing began'. Perhaps he has not quite done that. The ramifications of this 'great muddle' are very far-reaching indeed. But perhaps he has cleaned up enough of it to enable us to see the outlines of what may be called a working

[1] Cf. e.g. Carington's *Telepathy*, Chs. II and III, which contain numerous references to the relevant literature.

philosophy of Psychical Research. I say a 'working' philosophy, because I believe that the constructive theory which emerges is a somewhat rough-and-ready one, and leaves a lot of philosophical questions unanswered. All the same, it may be sufficient for our needs. It is certainly comprehensive, and it does introduce some sort of intelligible order into the whole field of 'supernormal' happenings; it also makes that relation to 'normal' ones a good deal less unintelligible than it was. The task of removing excrescences and tying up loose ends may be left to the professional philosophers, whose business it is.

This philosophy, Carington tells us, may be called Radical Positivism, though if left to himself he would have preferred to call it Factualism or Actualism.[1] His theory of Meaning is certainly positivistic and it is certainly radical. As he emphasizes himself, he is greatly indebted to such works as *The Meaning of Meaning*, by Mr. C. K. Ogden and Professor I. A. Richards, and to Professor A. J. Ayer's two books, *Language, Truth and Logic* and the *Foundations of Empirical Knowledge*. He accepts the positivistic theory of Meaning in its earliest and most uncompromising form. He holds, that is to say, that all sentences which have meaning fall into one or other of two classes: either they are verifiable by means of sensation or introspection; or else they are tautologous (i.e. true by definition but devoid of factual content) like the sentences of Formal Logic and Pure Mathematics. It follows that all metaphysical statements are meaningless. When a metaphysician tells us that the Absolute is not in time, or that there is a Pure Ego to which mental events belong, or that sensations are caused by Things-in-themselves which are for ever beyond the reach of observation, his remarks purport to be giving information about matters of fact. (They are certainly not tautologies, like the statements of Logic and Mathematics.) But they cannot possibly be verified by sensation or introspection.

[1] Ch. I, Sect. 8, p. 15.

Metaphysical statements, therefore, are not even false; they are just meaningless. Indeed, they are not really statements at all, though they have the grammatical form of statements. In short, they are nonsense.

But this anti-metaphysical part of Carington's argument is only a preliminary. The positive precept which emerges from it may be formulated in the words of Bertrand Russell, quoted at the head of Ch. IV: 'In dealing with any subject-matter, find out what entities are undeniably involved, and state everything in terms of these.'[1] This precept leads Carington, as it did Russell,[2] to adopt a form of the philosophy known as *Neutral Monism*. And that is what he offers us as a working philosophy of Psychical Research.

When we are discussing Matter, what are the entities undeniably involved? They are sense-data or appearances, what Carington prefers to call 'cognita' (because they are what we immediately cognize). And when we are discussing Mind, what are the entities undeniably involved? According to Neutral Monism, they are the very same ones: sense-data, including of course organic sense-data or 'bodily feelings', together with other entities which are basically of the same sort, namely images. If you examine what you call the contents of your mind at this moment, you will find nothing but a set of visual, tactual, and other 'environmental' sense-data, some bodily feelings and some images, among which there will probably be some images of words: just a set of cognita again, including some which are also constituents of the material world. That is what the contents of a mind are. And you will not find anything *but* its contents. (At any rate the Neutral Monist is sure that he cannot.) In particular, you will not find any entity—any mind-in-itself or ego-in-itself—which 'contains' them. But won't you find an act of 'being conscious' or 'being aware'? No. That is not introspectively verifiable. Consciousness is just a name

[1] *Our Knowledge of the External World* (Allen and Unwin, London), first edition p. 107. Russell points out that this maxim is equivalent to Occams' Razor. Cf. also Carington's remarks to the same effect in Ch. I, Sect. 9 (p. 17).

[2] See Russell's *Analysis of Mind* (Open Court Co., Chicago and London, 1914), and Carington's remarks about it at the end of Ch. VI.

for a property which a set of cognita have when they are suitably related *to each other.*[1] But will you not find an act of attending, or an act of willing perhaps, even if you find no act of awareness? Yes, if you like to put it so. But, according to the Neutral Monist, it is misleading to call them 'acts'; in so far as they are introspectively verifiable they are reducible to organic sensations, sensations of strain. Some emotions, then, at least? Yes, you may find them too; but they too (it would be maintained) are reducible to constellations of organic sensations, as the James-Lange theory pointed out. Hume has put the case better than anybody: 'When I enter most intimately into that I call *myself*, I always stumble upon some particular perception or other, of heat or cold, light or shade, love or hatred, pain or pleasure. I never can catch *myself* at any time without a perception, and never can observe anything but the perception.'[2]

Thus when we have seen through the nonsensical phraseology of the metaphysicians about material substances-in-themselves and minds-in-themselves, and look the bare facts in the face, we find that both Mind and Matter consist of neutral particulars, Carington's 'cognita' or Hume's 'perceptions'. Those particulars are in themselves neither mental nor material (hence the name '*Neutral* Monism'), but they are the constituents of mind and of matter alike. The words 'mind' and 'matter' stand merely for two different sorts of pattern or organization in which these same particulars are found to arrange themselves. In so far as they arrange themselves according to the laws of Physics, they are called 'material'. In so far as they arrange themselves according to the laws of Psychology, they are called 'mental' and Carington himself sometimes calls them 'psychons'.[3] But the very same set of particulars,

[1] Cf. Ch. V, Sect. 64, pp. 158–9.

[2] *Treatise of Human, Nature*, Bk. I, Pt. iv, Sect. 6, 'Of personal identity' (Everyman Edition, vol. I, p. 239). Hume's 'perceptions' are exactly equivalent to Carington's 'cognita'.

[3] This, it will be noticed, leaves open the theoretical possibility that they might on occasion arrange themselves in other quite different patterns as well, which would be neither material object nor yet minds as we ordinarily understand those words.

for example the set of colour-expanses which compose my present visual field, may be both mental and material at the same time. They are constituents of the several material objects which I am at present seeing—the paper, the pen, the table, the wall-paper, and so on—and they are also constituents of my mind. Sensation, as Russell says, is 'the intersection of mind and matter'.

What advantages are there in this queer system of thought, considered as a 'working philosophy' of Psychical Research? Whatever difficulties there may be in it from the point of view of the professional philosopher (and certainly he will find plenty) it does have one great virtue from the point of view of the psychical researcher. If I may say so, it gives him back his liberty, by dissolving away two kinds of *unity* which have become obstructive: the unity of the individual mind, and the unity of the individual material object. It seems perfectly clear that we can make no sense of supernormal mental phenomena—or indeed of abnormal ones either—so long as we insist on regarding the individual human mind, the *res cogitans* or 'thinking substance' of Descartes, as our fundamental unit. Whether the notion of 'mental substance' is sense or nonsense, at any rate it does not fit the empirical facts. The unity of a mind may be considered in two aspects, internal and external; and the essential point is that in both aspects it is a matter of degree, not a matter of all or none, as the Cartesian doctrine of mental substances would make it. Suppose there is an apparition which displays a certain amount of intelligence and purpose, but not very much. Shall we attribute an individual mind to it or not? Is the 'control' of an entranced medium an individual mind or not? Is a secondary personality, such as the celebrated Sally in the Sally Beauchamp case, an individual mind or not? Such questions are unanswerable as long as we insist on taking the individual mind as our fundamental unit. For then we have to take them as questions of all or none, which is just what they are not, and give a 'plain answer', yes or no, which is just what the facts do not admit of. To make any sense of the facts, we must change our unit, so to speak. We must take as our unit something much more elementary, the individual 'psychon'[1] as Carington called it, instead of the individual mind, and we must build up the various grades of mental entity out of these more elementary units: from not-very-purposive ghosts and Freudian complexes at the one end to the complete and healthily integrated human mind at the other, with mediumistic 'controls' and secondary personalities somewhere in the middle.

So much for the unity of a mind in its internal aspect. But in its external aspect also, the unity of a mind is a matter of degree. The facts of telepathy show that one mind is not cut off from another by any hard and fast boundary. Imagine two minds which are in continuous and unrestricted telepathic *rapport*.[2] Would there be any meaning in calling them two and not one? And if they are in partial and occasional telepathic *rapport* (which is all we find in fact) they are neither wholly identical nor wholly distinct. If we break them both up into systems of more elementary units we have a means of talking about such situations, which are very difficult to handle in the traditional 'substance' terminology. We can also hope to introduce some clarity into the obscure but useful notion of a common unconscious or subconscious, which many writers have postulated as the explanation of Telepathy.[3]

But of course the Neutral Monist philosophy dissolves the unity of the individual material object as well as the unity of the individual mind. It regards a material object not as a unitary substance, but as a 'logical construction' out of

[1] I should myself prefer the word 'idea', following Herbart. Of all the traditional philosophical psychologies, I suspect that Herbart's is the most illuminating to the student of Psychical Research.

[2] On the conception of telepathic *rapport*, see Carington's *Telepathy*, pp. 66–71.

[3] Cf. Ch. V, Sects. 80–1, pp. 207–14, below.

many different appearances (sense-data and/or sensibilia[1]). In the ordinary way appearances go together in those clumps or integrated groups which we commonly call 'Things'. But there is no *a priori* necessity about this. For all we can tell *a priori*, they are 'free to come and go "like the blessed gods" at their own sweet will', as Carington puts it (Sect. 49, p. 120), and he describes his own theory, in the same passage, as 'the Doctrine of Autonomous Appearances'. Accordingly, we might sometimes expect to find groups of appearances which are *less* complex than a complete and full-blooded 'Thing'. As a matter of fact we do: in hallucination,[2] for example; and again in such phenomena as mirror-images and mirages (both of which are quite complex systems of appearances, but less complex than a complete object); and most strikingly of all in the public or semi-public apparition[3] perceived by several observers at the same time, which has some of the properties of a material object but lacks others (for instance, it lasts only for a short time, is usually intangible and does not produce effects in other material objects which surround it). Perhaps we might also expect to find that appearances of *different* objects would sometimes show correlations with each other not explicable by the laws of Physics nor by chance-coincidence. In other words, there might be other forms of unity in the world of appearances, additional to, and cutting across, the familiar ones we call 'Things'. Whether we ever do find anything of the sort, I do not know. Superstitious people in all ages have supposed that they found them.

Here again, then, it seems that the Neutral Monist philosophy, by dissolving traditional unities which have become obstructive, and turning questions of all or none into questions of degree, may be able to give some useful guidance to

[1] In Carington's terminology 'cognita' and 'cognizables'.

[2] Cf. Carington's explanation of Precognition in App. III, by means of a 'dissociation' of the visual constituents of an object from its other constituents. In effect this turns Precognition into a kind of hallucination displaced in time. Cf. also pp. xix–xx, below.

[3] See G. N. M. Tyrrell, *Apparitions* (Myers Memorial Lecture, 1942. published by the Society for Psychical Research).

the psychical researcher. For Psychical Research, whether we like it or not, is concerned with the material world as well as the mental world, however dubious some of the 'physical phenomena' of the séance room may be.

But what of the interaction *between* matter and mind? Will Neutral Monism help us there? Carington evidently thought that it would. In particular, he thought it would throw light on the mysterious phenomenon of Psychokinesis, which had occupied a good deal of his attention in the last year or two of his life.[1] In Psychokinesis (P.K. for short) it would seem that mind causes changes in matter 'directly', i.e. independently of the ordinary mechanism of the nervous system and muscular apparatus. The reader must be warned that there are some distinguished psychical researchers who are not yet convinced that P.K. occurs at all. But there is some experimental evidence in favour of it,[2] and there is some evidence from 'spontaneous cases', e.g. from poltergeist phenomena. There are also of course the phenomena of physical mediumship. Many of them can no doubt be explained by fraud or malobservation; but it would be rash, I believe, to suppose that absolutely all of them can be explained in one or other of these ways. In short, the evidence for P.K. is strong enough to call for serious consideration; and we ought at least to ask ourselves what theory we should be obliged to hold, if the existing evidence were to be confirmed by further investigations. In any case, Carington himself was convinced of the reality of P.K., partly as a result of some experiments of his own; and he obviously did think that his Neutral Monist philosophy would provide a satisfactory explanation of it. The question is, how? Presumably that is what he planned to tell us in Ch. VI, which should therefore have been the most exciting and original part of the whole book. But as we have seen,

[1] At the end of Sect. 10 (p. 23) he promises to discuss it, and would presumably have done so in Ch. VI.

[2] Cf. Professor J. B. Rhine, *The Reach of the Mind* (Faber & Faber London, 1948), Chs. VI, VII, VIII.

Ch. VI was never finished. We have only a few fragmentary notes, and there is not a word in them about P.K.

It is not easy to guess what he might have said. But it may be worth while to make the attempt, however unsatisfactory the results may be. In making it, I have been guided by the following passages in the existing text: Sect. 10 *ad fin.*, p. 23, referred to just now; the concluding section, 58 of the chapter on Matter (pp. 43–5), since he tells us at the beginning of the existing fragment of Ch. VI (p. 229) that 'the basic answer to the question of how Mind and Matter are related, and, in principle, therefore, of how they may interact, was given at the end of Ch. IV'; the end of Sect. 48 (p. 117–18) especially the words 'with due reference to their possible dissociation or aberrancy'; and Appendix III on Precognition, which I take to be concerned with this possible dissociation or aberrancy of cognita. I am also drawing on my recollections, unfortunately very hazy, of various discussions I had with him during the last five or six years of his life.

Perhaps he might have argued on the following lines:—The chief problem about P.K. is not so much to make it intelligible but to remove the obstacles which make it appear *un*intelligible. The most important of these is precisely the metaphysical phraseology of substances and attributes whose nonsensical character has already been explained. So long as we think (or talk) of a mind as one sort of substance and a material object as another sort, even 'normal' mind-matter interaction, the interaction of mind and brain, appears to be unintelligible. A material substance is something extended in space and non-conscious; a mental substance is something conscious and unextended in space. It is supposed to be unintelligible that two entities so utterly different should interact with each other at all. In order to get out of this difficulty (which their own nonsensical terminology has got them into) metaphysicians have had recourse to the most extraordinary expedients: Psycho-physical Parallelism—with or without Divine Assistance—which retains the two substances but denies the interaction; and Materialism and Mentalism, which deny the existence of one or other of the two substances, and solve the problem by abolishing the relevant empirical facts.

But if we adopt a Neutral Monist philosophy (so Carington might have argued) the whole difficulty vanishes. As he puts it in Ch. I, where he is giving an outline of the main points he intended to make,[1] 'Mind and Matter are not two qualitatively different "orders of reality" or the like, but inseparably interlocking and interwoven organizations or patterns of the neutral and irreducible entities or 'particulars' (cognizables) of which the universe is composed'. Thus we no longer have to ask how one substance can interact with another very different one. We no longer worry ourselves with 'substances' at all. Instead, we simply observe the correlations which are in fact found to hold between the neutral particulars which are the ultimate constituents of both the material and the mental worlds alike. The visual 'cognitum' of a pin pricking my finger is regularly followed by a 'cognitum' of pain. The cognitum which we describe as 'an image of the movement of my hand' is regularly followed by the visual and other cognita constituting the actual physical movement. And if it be still objected that such correlations are not intelligible, even though undeniable as brute facts, I think Carington would be entitled to reply by asking the objector what he means by 'intelligible'. Behind this demand for intelligibility (so Carington might have argued) there lies another metaphysical muddle: the one exposed long ago by Hume, the confusion between causation and logical entailment.[2]

What bearing have these considerations on P.K.? Simply this: they remove the intellectual inhibitions which might

[1] Ch. I, Sect. 10, 'Course of the Discussion', point 9 (p. 22).

[2] On the notion of 'causal necessity', cf. Ch. IV, Sect. 44, p. 106. But in that passage Carington does not distinguish between two senses of the word 'necessity': the *compulsion* sense, and the *logical entailment* sense. On his principles, both of them are equally 'metaphysical'. And Hume exposed them both.

prevent us from accepting the empirical facts, on the ground that such things 'can't happen'. Actually there is no *a priori* reason whatever why P.K. should not happen. Whether it does happen or not is just a question of fact; and Carington himself, as we have seen, was satisfied that it does. The *a priori* arguments, if there were any force in them, would show that even the most 'normal' mind-matter interactions are impossible. If there is no force in them, as we have seen there is not, the way is left clear for accepting *all* the forms of psycho-physical interaction—normal or supernormal—which are empirically found to occur.

This, or something like this, is what I imagine Carington intended to say. But if he had said no more than this, I think that his readers would still have felt unsatisfied. And I imagine that he did intend to say something more, though it is difficult to guess what it would have been. He would have admitted, I think, that there is a legitimate scientific sense of the word 'intelligible' (as opposed to the illegitimate metaphysical sense) in which P.K. does appear *prima facie* to be unintelligible. Allowing that it does occur and that there is no *a priori* reason why it should not, one might still complain that it is entirely without analogy to all the other causal transactions, or regularities, which we know of: a mere oddity, which we must accept with natural piety if the evidence compels us to do so, but which does not 'fit in anywhere' among the other facts we know. I believe that Carington would have tried to show that it does after all 'fit in somewhere' and does have some analogy with other known facts. How he would have set about it, I can only conjecture. But on the face of it there are two lines of thought which might possibly be helpful. Both are consistent with Carington's Neutral Monist principles, and the second is directly suggested by them.

The first is to generalize the notion of what psychologists call *ideo-motor action.*[1] In the phenomenon of ideo-motor

[1] Cf. William James, *Principles of Psychology*, Vol. 2, pp. 522 ff.

action, we find that an idea tends to fulfil itself or execute itself automatically through the muscular apparatus of the body, and will in fact do so unless other ideas are present to inhibit it. One might suggest that this operation is merely a special case of something wider: namely a tendency which ideas have to 'realize' themselves in the material world in any way they can. Someone has remarked that in the phenomena of physical mediumship 'we see the subjective walking about the room': that is, ideas in the medium's unconscious embody themselves in a visually-perceptible form. And in poltergeist phenomena we seem to see the same.

If this suggestion seems too queer—though it is not much queerer than the phenomenon which has to be explained—we may consider another one which Carington himself might perhaps have preferred. We may approach P.K. through a consideration of *hallucination*. We are inclined to say that a hallucinatory entity is something 'purely mental', as opposed to a 'real object' which is physical. But according to Carington's philosophy the difference between the hallucinatory and the physically real is in the end only a difference of degree. A hallucinatory entity, the celebrated pink rat for instance, is composed of sense-data or appearances (cognita) just as a 'real' object is. What is wrong with it, what inclines us to call it 'unreal', is the fact that there are not *enough* of them. For example, the hallucinatory rat can be seen from the front but not from the back; it is visible but not tangible; it can be perceived by one percipient but not by more; and it endures only for a minute or so. But some hallucinations do better than this. Apparitions, for example, are sometimes public to several percipients, can be seen from several different points of view, and endure for considerable periods of time—though not as long as they would if they were 'real' human beings. Now suppose there was an apparition which had unrestricted publicity, i.e. was public to an indefinite number of points of view and an indefinite number of observers: suppose that there are tangible as well as visible particulars among the appearances (or 'cognita') which are its constituents; suppose it endures for half an hour and then disappears. We should not know whether to call it an unusually prolonged and complex hallucination, or a very queer 'real object' (queer, because we should not know how it got into the room, or why it abruptly vanished into thin air). In point of fact, it would be something intermediate between the two: a complex system of 'cognita', but not quite complex enough to count as a complete material object.

But what have hallucinations to do with P.K., the reader may ask. The connexion is this. Hallucinatory entities are produced by *mental* causes: either by some idea in the unconscious of the percipient himself; or, if it is a telepathic hallucination, by an idea in the mind (conscious or more likely subconscious) of someone else. The study of hallucinations shows us that a set of particulars which is, as it were, a very incomplete material object can be produced by purely mental causes; and the study of apparitions shows that something much nearer to a complete material object can sometimes be so produced. Now imagine this process pushed to the limit, so to speak. We might expect that occasionally a complete material object or a complete physical event would be produced by purely mental causes. And that, perhaps, is what happens in P.K.

This explanation too looks like the wildest nonsense. But Carington's Neutral Monist philosophy does at least make it appear a little less nonsensical than it would be if the traditional dualistic metaphysics of mental and material substances were retained.

H. H. PRICE

If anything in the world is worth wishing for . . . it is that a ray of light should fall on the obscurity of our being, and that we should gain some explanation of our mysterious existence, in which nothing is clear but misery and vanity.

SCHOPENHAUER

This won't do. Our business is to bring order, proportion, light into what is happening. That where there has been falsehood and muddle there shall be knowledge and clearness, conception for misconception. . . . Get on with it.

MARY BUTTS

Of what huge devils hid the stars,
Yet fell at a pistol-flash.

CHESTERTON

I

OUTLINE OF THE DISCUSSION

> Philosophy, from the earliest times, has made greater claims, and achieved fewer results, than any other branch of learning.
>
> BERTRAND RUSSELL, *Our Knowledge of the External World*

1. *Object of the Work.* The object of this book is to clean up once and for all—in principle and in outline at least—this muddle about the relation of Mind to Matter, which has fretted philosophers ever since philosophizing began.

In these preliminary remarks I propose to indicate, briefly and colloquially, first, what the problem is about, second, why it is so important, third, why the belief that it can be solved is not necessarily so preposterous as it may seem; and finally, the general course that the discussion will take, together with a few notes on the guiding principles adopted, and the special nature of the difficulties—such as they are—likely to confront the reader.

As regards the first: To state the problem fully and exactly—that is to say, to expand the terms 'Matter' and 'Mind' to the point at which there can be no possible ambiguity as to what we are talking about—would be to solve it, as we shall see; so for the moment the following few words must suffice to make clear the general nature of the topic.

From time immemorial, man seems to have had a more or less innate conviction that there is a sharp distinction to be drawn between two importantly different 'orders of existence'—the phrase will serve well enough for the moment—to which he purports to refer by such contrasted terms as 'Matter' on the one hand, and 'Mind' or 'Consciousness' on the other. The question how one order is related to the other has been, at all stages of thinking, a favourite topic of speculation, especially among philosophers and metaphysicians. Floods of ink have been spilt, and controversy has run high, in debating whether it is correct to say that Matter, or Mind, as the case may be, is the only 'Reality', with the other no more than some sort of an illusory or 'epiphenomenal' (whatever, if anything, this may mean) poor relation thereof—sooner or later to be relegated to the lethal chamber, or at most pensioned off as an unimportant dependent; or whether both are 'Real', though in different modes, or both alike. Aspects of some Underlying Reality in some way more real and enduring than they.

The average man of to-day takes, I suppose, little more interest in these speculations than did his forefathers, though I think he ought to, and on no account be content to 'leave it to the philosophers'; but I think it probable that, in his heart of hearts at least and however much he may have been bludgeoned into rendering lip-service to the materialists, he still recognizes the distinction and still believes it to be fundamental to the nature of things. We are all of us, he would contend, perfectly familiar with various objects—stones and water, trees and plants, the bodies of other animals and our own—which we speak of as 'material' and refer to collectively as 'matter'; but also with what he would call 'thoughts' and 'imaginings' (and he would probably include 'willings' and 'desires' and 'memories') which we call 'mental', or concerned with the 'mind', and are, on the face of them, radically different from 'matter' and 'material' objects.

It is, roughly, with the justification for this distinction, and with the 'ultimate nature'—again an imperfect phrase must serve—of the relation between these two 'orders' that the present work is concerned.

2. *Importance of the Subject.* In view of the fact that the subject has been left almost entirely to philosophers, and the way in which discussion of it degenerates into interminable and unedifying wrangles about words, the average reader might be forgiven for supposing that it is purely of academic interest, and of no more practical application than the problem, say, of how 'redness' is related to 'circularity'.

I am anxious to insist from the start that this is not the case; but that, on the contrary, the matter is one of supreme and immediate practical importance on any but the most superficial view—at least on the assumption that the continued existence, well-being, and happiness of mankind is to be regarded as practically important.

It needs no words of mine to emphasize that the world to-day is in a dreadful state, or that it shows little prospect of speedy improvement, for that seems to be fairly generally admitted; but there does still seem to be some doubt as to where the root of the trouble lies. Very briefly, the basic problem is not technical but moral—as, of course, has been remarked before. Fundamentally, the trouble is not that the scientists have discovered the inner secrets of the atom and have devised means of blowing our vaunted civilization to smithereens in a few hours; or even, more generally, that we have developed powers of all kinds undreamed of even a couple of generations ago. There is no kind of necessity that requires us to use our knowledge in destructive ways, and every sort of reason why we should not; yet everyone is scared out of his wits—and with good cause—that this is exactly what we shall soon find ourselves doing. The basic trouble is that there does not exist in the world to-day any set of moral principles, or theory of living, or philosophy—or 'religion', if you will—common to all mankind, and as universally accepted as the laws of physics, that indicates unequivocally what we ought to do and why.

Conflicting ideologies of economic and political character have taken the place, for public purposes at least, of religions in the older sense; but all such systems rest in the last resort, no less than the religions did, on certain assumptions—true or false, expressed or implied—about the nature of man, and on deductions from these as to what is 'good for' him and what he 'ought' to do or have done to him. If our views on these matters are erroneous, and man is something other than we have assumed, then we are pretty well bound, it seems to me, to go wrong and end in disaster; for we shall be in the position of men seeking to devise a machine, or build a bridge, not knowing the properties of the materials to be used. But if they are correct, there is at least a reasonable chance of being able to design a social system that will work well—unless, of course, we destroy ourselves in quarrels over the best way of doing that on which we all agree.

But of all questions relevant to the nature of man, that of the relation of Mind to Matter is surely the most fundamental. There is no need to work out implications in detail here; but it does seem at least highly probable that our attitude towards our fellow man—which is what counts in the long run—is likely to depend very considerably on whether we regard him (and ourselves) as merely a highly complicated automaton to which 'mind' and 'consciousness' are but trifling incidentals, or (to take the other extreme) as essentially an Immortal Soul (supposing this to mean something) who happens to be encumbered with a body on which he can exert mysterious 'forces' of Will and Choice and Decision. I submit that, until we have cleared up this problem of Mind and Matter—of 'the Mind and its Place in Nature', as Professor Broad has it[1]—and have done so beyond any possibility of argument, we cannot hope to evolve that agreed philosophy or religion or code of ethics which we so desperately need at the present time, and without which we seem at least as likely as not to perish miserably and soon. But we'd better be quick about it.

The relation of Mind to Matter is also, of course, vital to the whole question of whether a man's conscious existence terminates altogether with the death of his body, or whether in any but the most Pickwickian sense he may reasonably be said to survive it; and to many people this will appear to be of the highest importance. But although this is an issue which happens to have occupied my own attention over more years than I care to think about,[2] and one which I hope to deal with more fully in another place, I cannot but feel that, except in so far as it bears upon the situation I have just been discussing, it is of quite secondary importance; and it is not on this account that I invite the reader to follow me in what I have to say.

[1] C. D. Broad, *The Mind and its Place in Nature* (Kegan Paul, London, 1925).

[2] Cf. *Telepathy* by W. Whately Carington (Methuen, London, 1945), Ch. XI.

3. *Possibility of Solution.* The claim implied in my opening sentence, to the effect that I can succeed in a task which has baffled so many great, wise, and eminent men, may well savour of excessive optimism, and even appear to convict me, despite the very considerable diffidence that I feel, of a presumption tantamount to impertinence; especially as I am in no sense a professional philosopher. On the face of it the implication is that, if I can do what previous thinkers could not, then I must be cleverer or more intelligent than they; and this is certainly very far from true.

The personal aspect is, of course, trivial, though I would naturally prefer to avoid any such impeachment; but I think it important that the reader should not start off with the conviction that a problem which has baffled the greatest intellects of humanity for more than two thousand years is almost certain to be insoluble, so that it is waste of time even to consider it; for this would be bound to increase his resistance (likely in any event to be strong) to accepting the thesis I am about to propound.

As a matter of fact, any reader who feels like this is very nearly right; for the problem *is* inherently insoluble by the methods which metaphysicians (in whose province it has always been deemed to lie) have invariably adopted. If I were proposing to do the same kind of thing as they, only more so, I should be in no better position than a man proposing a new and more complicated tackle for hoisting himself by his own boot-laces. For the method of the metaphysician has always been to attempt to deduce conclusions about the factual universe, in which material and mental phenomena are observable, from *a priori* premises; and it has now been made clear (not by me)[1] that this is as flatly impossible as squaring the circle (on which, incidentally, a great many mathematicians of eminence wasted a great deal of time) or trying to detect absolute motion.

This has come about through the work of various logicians

[1] Cf. among other works Ayer's *Language, Truth and Logic* (Gollancz, London, 1938; second edition, 1947), to which I am greatly indebted in this connexion.

who have shown the falsity of the belief (current at least up to the time of Kant[1] and still lingering) that there are propositions about matters of fact (e.g. geometrical axioms, etc.) which can be known 'intuitively' to be true; whereas all true propositions must be either tautological or empirical.

Other work, on the subject of Meaning, notably that of Ogden and Richards,[2] to which I shall refer extensively in due course, has elucidated the rationale of the average reader's private conviction that the writings of metaphysicians are just so much meaningless nonsense, which may profitably be consigned to the waste-paper basket without further ado.[3] And certain psychological considerations, to be briefly indicated below, make it clear how it was almost inevitable that men of such ability should fall as they did into the linguistic pitfalls that beset them.

My advantage over my predecessors, in fact, is not that of superior intelligence or ability (presumably much the contrary), but that I happen to be in a position to see *why* they were bound to fail, and they were not.

All that I am doing in what follows is to insist on the acceptance of certain plain facts—so simple and obvious, for the most part, as to escape notice altogether—and on refusing to be lured aside by siren-songs of unanalysed words, which are found on closer scrutiny not to refer to anything whatever.

4. *Current Views on Matter and Mind.* For the last century or thereabouts, the physical scientists have ruled the intellectual roost; and not unnaturally, in view of their spectacular achievements in the study and exploitation of the properties of matter, and—from the logical point of view particularly—their successes in predicting the course of observable phenomena.

Their work has led to a host of discoveries and inventions, in agriculture, industry, medicine, and so forth, which have greatly ameliorated human life; and again and again their

[1] Cf. Ayer, loc. cit., Ch. IV.

[2] C. K. Ogden and I. A. Richards, *The Meaning of Meaning* (Kegan Paul, London, fourth edition, 1936).

[3] Except, of course, in so far as they may happen to contain advances in pure logic, having nothing to do with the subject-matter discussed.

pronouncements have proved correct. It is accordingly not surprising that their opinions have been conceded an almost priestly authority, even on matters not strictly within their field, or when not logically entailed by the discoveries, etc., made.

There is, of course, no official scientific creed, as one might speak of the Apostles' or Nicene creeds in connexion with the Christian religion; but, although I doubt whether even the most hardened materialist would deny consciousness as a fact (i.e. that he himself was conscious), I think most scientists—at any rate if speaking in at all an official capacity—would adopt one or other of two views. They would say, I think, either that the causal relation between Matter and Mind is purely a one-way affair, so that changes in matter may cause changes in mind (consciousness), but not reversely; or else that the two categories of phenomena, material and mental, are causally altogether disparate and to be regarded as in wholly watertight compartments incapable of inter-acting with each other. The first view would be describable, I suppose, as Materialistic Monism, the second perhaps as Radical Dualism.[1]

What I have just said may well be criticized as incomplete, or even in some degree inaccurate; but this is of no great importance here, for there is no doubt that the traditional policy of physical scientists (including physiologists and many psychologists) has been systematically to ignore 'mind', 'consciousness', 'spirit', etc., as possible causal factors in the phenomena they study, and thus, in process of time, virtually to deny their 'real' or 'effective' existence. And it is important to understand that, if this policy had not been adopted, scientific progress would have been almost impossible or at least enormously handicapped.

In the pre-scientific era human thought was largely dominated by superstition (by no means yet wholly dead, as reference to the daily press will show), and the universe was supposedly populated by innumerable spirits, transcendental forces, planetary Influences, and animistic deities of all

[1] The best known form of the second view is Psychophysical Parallelism. (H. H. P.)

kinds—all of which were believed to have power over mundane events of every description. But it is manifestly a hopeless task to undertake the systematic study of material phenomena, or the discovery of the laws which govern them (i.e. the observable regularities from which their course can be predicted) if you believe that whatever laws there may be are liable to be overridden at any moment by the whims of a malevolent spirit, the vagaries of the will of God, or the muttering of incantations by the local magician.

But to free oneself from such beliefs—which, it must be remembered, were in those days as much taken for granted as the air one breathed—and thereby to secure, so to say, a clear field in which to work, must have been very much more difficult and have called for far greater strength of mind than it is easy for us to realize to-day. It could be, and of course occasionally was, achieved by the elect few; but for such emancipation to occur on the large scale that extensive scientific work demanded, it was necessary to find a justifying philosophy, that is to say, to rationalize the disbelief, and to do so, if possible, in such a way as not too grossly to offend existing Authority.

5. *Cartesian Dualism.* It has been pointed out to me by Professor Price that the categorical separation of Matter from Mind was largely due to the influence of Descartes, according to whose system of philosophy the distinction between 'thought and extension' (i.e. mind and matter) was absolute —'so absolute that only the continual interference of God could account for their harmony'.[1] This doctrine, which the great prestige of Descartes as a mathematician did much to spread, was evidently just what was needed. By asserting the complete causal independence of Matter and Mind, it gave the scientist the free field he needed; it did not deny the reality of mind or consciousness, as would have been contrary to common sense and universal experience; nor did it deny the manifest parallelism of mental and physical phenomena (e.g. volition and action) which would have been equally so, but it did deny that this parallelism was the result of causal

[1] Wording from *Webster's Dictionary*, heading *Cartesianism.*

interaction, and thereby freed the scientist from paying any attention to it; moreover, by attributing it to 'the continual interference of God' it doubtless succeeded in placating the contemporary theologians. In fact, it was 'made to measure', so to say, as a scientific creed.

There was, of course, one trifling difficulty about it, namely the extremely important part played by 'God', for Descartes would have been the last man to contend that the observed parallelism could be ascribed to chance alone, nor would any scientist support such a view. For a long time, no doubt, most scientists remained more or less religious believers, to the extent at any rate that they had no great difficulty in accepting the notion of this benevolent activity on the part of Deity—*ex hypothesi* it made no difference to the material phenomena they studied. But later they rightly found occasion to challenge the conception of 'God' defined in some way other than as 'that which is responsible for psycho-physical parallelism'; for example, when defined as 'He who brings rain in response to prayer'. The anthropomorphic deity of jealous, punitive, flatterable, etc., attributes was very properly rejected; but with him disappeared also the Ensurer of Parallelism.

As this process took place, the observed facts of parallelism were left more and more in the air, unsupported by any explanation—until some misdirected genius[1] coined the blessed word 'epiphenomenalism'. This is merely a label for the doctrine that consciousness or mental events, etc., are caused by (i.e. are observed with a certain regularity to accompany or follow) physical events (brain processes, etc.), but are incapable of exerting any influence upon them. The first part of this we knew already, and our understanding is not advanced by giving the facts a new name; the second is pure dogma, and, as we shall see later, almost certainly untrue.

However, though the logic of the Cartesian dichotomy was faulty, and the attempts to patch it up no better, the attitude adopted, considered as a practical *policy*, has been triumphantly vindicated. Throughout an immense range of studies it has been found in practice that to ignore the possibility of

[1] It was, I believe, T. H. Huxley. (H.H.P.)

physical phenomena being influenced by mental, etc., factors does not lead to discernible discrepancies—that is to say, it has been found safe to assume that no such influence occurs—while such small exceptions as may have been noted have successfully been explained away or attributed to experimental error.[1] Thus the physicists have been well justified in concluding that the assumption is true for the field within which their investigations have been conducted, and to the accuracy of the measurements, etc., employed. But it does not in the least follow that it is necessarily true for different phenomena or for more accurate measurements.

There are tolerably close parallels to this within the domain of physical science itself. We do not take account of the influence of electrostatic forces when calculating the flight of a projectile, or demonstrating the principles of mechanics with an Atwood's machine, even though we know very well that friction between any two different substances invariably produces an electrostatic effect in some degree. We say, rightly, that these forces are negligible in comparison with those in which we are interested; yet, in certain circumstances, they may be very important, as when the friction of hydrogen escaping through the valves of an airship may produce sparks capable of igniting the whole mass, or, of course, more familiarly in the case of lightning. But we do not on this account deny the 'reality' of electricity or its 'causal' connexion with gross matter, though all electrical science grew out of the feeblest effects of amber-rubbing, etc.

A more cautious attitude on the part of physical scientists, and one more strictly logical than that actually adopted, would have been to rest content with saying that mental factors, and 'mind' generally, very seldom (if ever) exert more than a negligible influence on the course of physical phenomena; rather than committing themselves to the view—as to all intents and purposes they have done—that any such influence is categorically impossible.

[1] Or, of course, flatly—and by no means always unjustifiably—denied, as in the case of the alleged 'physical phenomena of spiritualism', for which (I hasten to add) I hold no brief here.

6. *Philosophers and Philosophy.* I have already indicated that it is part and parcel of my thesis that all Metaphysics is nonsense, and I shall devote the next chapter to explaining more fully why this is so. But it would not be conducive to the proper attitude of approach to suppose that by maintaining this I am thereby seeking to disparage the great philosophers, or to underrate the services they have rendered to humanity.

Metaphysics, said to be 'That division of philosophy which includes ontology, or the science of being, and cosmology, or the science of the fundamental causes and processes in things' (*Webster's Dictionary*), is not the only activity in which philosophers have engaged in their professional capacity. There are also Ethics and Logic.

It is not yet clear to me whether the logical status of traditional Ethics is in any better case than that of Metaphysics—I think not; but there can be no doubt that philosophers have in the main tendered sage advice to the successive generations of mankind, even though the alleged 'proofs' of its validity may have been uniformly in error. It is, indeed, not merely possible, but almost standard practice, to support perfectly sound conclusions by completely preposterous arguments, and even to believe that the arguments not only guarantee the conclusions but are the sole means of arriving at them; whereas what usually happens is that the conclusions are formed first, crystallizing as it were out of a concentrated solution of assorted experience, and the arguments are invented afterwards as a kind of pious tribute to the Goddess of Reason. But this does not affect the validity of the conclusions themselves; and if a philosopher teaches that a man's wealth consists mainly in the abundance of the things he can do without, the aphorism is no less wise and valuable because it is ostensibly 'proved' by a lot of meaningless verbiage about appearance and reality and eternal values, instead of being frankly based, in Kipling's phrase, on 'extended observation of the ways and works of man'.

As for Logic, practically the whole of this all-important subject other than pure mathematics (and a very fair amount of this also) has been worked out by philosophers, who have thereby placed a tool of inestimable value in the hands of mankind. But it is important to realize that it *is* a tool and not a product. It operates by the manipulation of symbols (verbal or mathematical) in analytic (tautological) propositions, the truth of which is assured (if the manipulations are correctly performed) by the way in which the symbols are defined to start with. And just as the most perfect sausage machine cannot produce sausages of higher quality than the raw material fed into it—as who does not poignantly realize in these days of breadcrumbs and soya—so even the most perfect logician cannot produce conclusions more applicable to the factual world than are the definitions and axioms with which he starts. *If* there be existents which conform to the definitions he uses, then his results apply; but if not, then not. Many a philosopher, I suspect, has excogitated logical operations of novelty and value while indulging his passion for metaphysical speculation, just as many chemical processes of practical utility were doubtless discovered by the alchemists in their search for the elixir of life or the universal solvent—or geometrical constructions by would-be circle squarers.[1]

But apart from these technical services, it is to be noted that the communication of veridical propositions is not the only valuable function of language; it may also have hortatory, stimulative, emotional, or purely aesthetic virtues. Plato's famous image of the prisoners in the cave,[2] for example, has a purely literary value, without which the world would have been the poorer, which has nothing to do with its metaphysical implications, which are as nonsensical as any others; and his *Republic* as a whole (in which it occurs), whether we agree or not either with his conclusions or his methods of reaching them, must have inspired thousands with the idea

[1] It has been suggested that this affords a justification of metaphysics, and also that a metaphysician might happen to hit upon a whole system later found to fit the world of fact, as has sometimes happened in mathematics (e.g. the Riemann-Christoffel Tensor used by Einstein); but, methodologically speaking, this seems to me altogether too suggestive of Huxley's hypothetical monkeys hammering typewriters.

[2] *Republic*, Bk. VII, *ad init.*

that human society might be rationally planned for the benefit of its members, instead of just struggling along at the whims of tyrants.

Spinoza again, though an archetypal metaphysician, can hardly be read without profit and a certain uplifting of the soul. But Spinoza was a very great man, whose thoughts about what he called 'God' were so far in advance of his age (and for the most part of ours also) that he was promptly denounced as an atheist; and he probably could not have written a treatise on sewage analysis without infusing it with his own austere nobility.

And, of course, in somewhat lighter vein our friend Hegel is good for a hearty laugh on almost every page.

Broadly speaking, then, humanity is greatly in the philosophers' debt, though not usually for the reasons they would probably have claimed. In the main they have been on the side of the angels, and have been second only, if at all, to the great philanthropists in their concern for the long-term happiness of mankind, and to the greatest scientists in their desire to let in the light. And it is hardly to be counted against them, as we shall see in a moment, that they have wasted so much of their time trying to unravel knots in endless cords instead of slashing them across with an Alexandrine cutlass.

So if I speak disparagingly of philosophers and their work, it must be clearly understood that it is only in their capacity as metaphysicians that I do so.

7. *The Handicaps of Philosophers.* We shall deal in the next chapter with the general topic of why metaphysics is foredoomed to futility; but it is worth noting here, slightly anticipating points which will be emphasized later, that there is one very serious handicap under which philosophers have always and inevitably laboured.

I do not refer here to the rather obvious disadvantages of being credited with wisdom, or at least with learning, above the average, so that it is difficult to say flatly 'I don't know' without loss of 'face'; but to the somewhat subtler point that a philosopher is very much in the unenviable position of a trap-nested hen.

Whenever he lays a good, sound, hatchable egg in the form of a meaningful statement of verifiable character about a matter of observable fact, and it is verified—as of course has often happened—it is automatically and *ipso facto* snatched away from him and at once incorporated in one or other of the special sciences. Even if it be found false, he has still added, albeit in a negative sense, to the sum total of knowledge in that science; and even if it cannot be immediately tested, but could be under suitable conditions (e.g. if he declares that the far side of the moon is puce) it will still rank as a scientific hypothesis.

If, on the other hand, the egg takes the form of a worthwhile pronouncement about the use of symbols, then it is similarly snatched away and incorporated in the appropriate branch of logic or mathematics.

But philosophers, in the words of the cliché, 'are only human', and presumably, like the rest of us, enjoy the feeling that they are experts and specialists in a field on which lesser mortals are not entitled to intrude. To ensure the possession of such a field as their private preserve they must clearly, in view of what I have just said, specialize in remarks which are not purely about symbols and yet are not susceptible of being put to factual test, else they will be bereft of them in the one way or the other. Their remarks (propositions) must accordingly either consist of pure gibberish (and not many have fallen quite so low as this), or else they must purport and appear to be about matters of fact (e.g. the Nature of God, the Attributes of Reality, etc.) and yet be incapable of being verified or refuted.[1]

Unfortunately, propositions of this kind, with which the pages of the metaphysicians are crammed, are of necessity completely meaningless, nonsensical, and void of literal significance. This is because any such proposition is a complex symbol purporting to refer to a complex 'object', so to call it provisionally (i.e. some sort of event or situation), which is

[1] More accurately, of such a nature that it is inherently impossible to obtain evidence or make observations bearing on the probability of their truth or falsity. Cf. Ayer, *Language, Truth and Logic*, pp. 22–6.

for practical purposes its meaning; and if, as is the case *ex hypothesi*, it is inherently impossible to identify the situation in the universe, therefore to ascertain whether the proposition is true or not, it might just as well not have been uttered, which is another way of saying that it is meaningless. This is a trifle loose, but I shall tighten up our ideas about Meaning in my third chapter.

8. *Attitude of this Work: Radical Positivism.* A few words will not be out of place regarding the line of approach and guiding principles of this work. I have been variously told by those who ought to know that I am a Phenomenalist, a Neutral Monist, possibly a Sensationalist—not to mention, of course, many other things usually more opprobrious but not relevant to the present context. Personally, I have always been less interested in the labels on bottles than in the wine inside, and I am not greatly concerned to accept or repudiate any of these characterizations, none of which, so far as my limited knowledge of these schools goes, seems to me wholly applicable.

Left to myself, I think I should say that I am, or try to be, a Factualist, or maybe an Actualist, on the ground that it seems to me to be all-important to stick with ferocious resolution to observable fact, and to ask oneself again and again what is actually going on—boiled down, so to say, to irreducible terms—rather than be lured into following up what may be purely verbal trails into labyrinths thick with linguistic pitfalls. Indeed, my principal insistence is on breaking down what would ordinarily be regarded as hard and 'atomic' statements of fact into the irreducible constituents of what we are actually aware of, or do immediately know (cognize) on the occasions in question.[1]

But if something rather more academic than this be demanded, I think I would choose the phrase 'Radical Positivist', though I might have preferred 'Radical Empiricist' if William James had not appropriated it years ago for a somewhat different doctrine of his own.

[1] Note that the term 'observable fact' is not synonymous with 'material object' or 'material event'. An hallucination or, as we would say, a purely imaginary object, is a perfectly good observable fact, even though there be no material object present of the kind we imagine or 'think we see'.

Under the heading 'Positivism' Webster says, 'A system of philosophy originated by Auguste Comte. It excludes . . . everything but the natural phenomena or properties of knowable things, together with their invariable relations of coexistence and succession as occurring in time and space. All other types of explanation are repudiated as "theological" or "metaphysical".'

This gives the general idea well enough, so far as it goes; but I would press the principle very much farther. In the first place, the attitude I want to see adopted and unswervingly applied would be of no value if all it did was to lead to just another of these 'systems of philosophy', accepted only by those who happen to feel like it, and as open to all the rest to cavillings by those who do not. Either the way of going about thinking that you adopt is the only right one, or it is nonsense. Either you are going to confine yourself in your last analyses to statements exclusively in terms of what is immediately and indubitably knowable, or you are not; and if you are not, then you cannot set any limit to the nature or number of the hypothetical, mythological, etc., entities (alleged) that you are going to admit.

Secondly, I will have no truck whatever with 'phenomena and properties' *of* knowable[1] *things*. Phenomena and properties, if by these terms are meant nothing more than observables or cognizables, yes; but 'things' which they are said to be 'of'—a thousand times *No*! I shall insist at length on this later; at the moment I will only say that a so-called 'positivism' that accepts 'things' said to be knowable, and properties said to be 'of' them, is not *radical*, but only a half-way house.

Nor am I altogether at home with the Logical Positivists: not that I have any objection to the logic, while there is much of their work, especially that of Ayer, which I greatly admire. On the other hand, they seem occasionally to come to very odd conclusions (as, for example, Carnap's 'self-consistency'

[1] One might have expected the author to say 'unknowable' in view of his arguments below (pp. 107–9). Presumably he means 'allegedly knowable'. (H. H. P.)

criterion of the truth of propositions, criticized by Ayer[1] himself); and they do not understand Meaning. Above all, perhaps, they appear to me to lack the courage, as do so many others, to drive their thought through to the limit, for fear (as I suspect) of offending the physicists and of being forced to some conclusion that physicists regard as distasteful.

It is very odd, if a minor digression may be pardoned, how often the most brilliant and even the most resolute thinkers seem to swerve aside just when their thought is getting really exciting. It is as if the Truth (if I may be pardoned the expression for a moment) which they purport to be seeking were a kind of Pandora's box, almost certain to release something alarming and unpleasant to anyone who raises the lid. Even Bertrand Russell, for example, the greatest philosopher of our day, than whom there can be few living men more morally courageous, seems always to be afraid of discovering that, after all, there *may* be some sort of a God in the box; and Eddington[2] that there may, after all, be *no* sort of God in the box. Most timorous of all, perhaps, are those who most stoutly declare that there is *nothing* in the box, but are no keener than the rest on turning the key that opens it.

9. *Guiding Principles.* Speaking at a more or less colloquial level, I consider that the guiding principles which should regulate all attempts at serious thinking form a kind of extension or intensification of the famous maxim known as Occam's Razor, viz. *Entia praeter necessitatem non multiplicanda sunt*, literally, 'Entities are not to be multiplied beyond necessity'; or, approximately, 'You must not introduce into your explanations, etc., more basic conceptions than are necessary for the purpose'; more informally, or 'Boil everything down to terms of as few irreducibles as possible'.

It seems to me imperative to conform to the following

[1] *Foundations of Empirical Knowledge* (Macmillan, London, 1940), Ch. II, Sect. 9.

[2] It is perhaps hardly fair to count Eddington as a philosopher, since he was primarily an astronomer and mathematician; but his would-be philosophical writings have probably exercised more influence on the thinking laity in the course of the last twenty years than all the professional philosophers put together.

rules, if confusion and the talking of nonsense are to be avoided:

1. It is futile, time-wasting, and therefore illegitimate, to use words, make statements, or formulate hypotheses, if you do not know what you mean by them; that is to say, if you are not quite clear as to what existents you are referring to when you use the words, etc.; i.e. if you literally do not know *what* you are talking about.

2. It is no use saying that *you* know what they mean, if you are not prepared, in principle, to make their meaning completely and unambiguously clear to others; that is to say, to take steps to ensure that when they use the words, or hear you use them, they refer to the same existents that you do; i.e. that *they* know what you are talking about. If not, then, so far as they are concerned, you are literally talking nonsense.

3. It is futile, time-wasting, and therefore illegitimate, to distinguish 'in theory', i.e. for purposes of verbal wrangling between two or more statements or hypotheses, particularly between any statement and its contradictory, if you are not prepared to indicate, in principle, how they are to be distinguished 'in practice'; that is to say, what difference in terms of observation or experience the alleged distinction will make; or what observations, experiences, etc., would be relevant to deciding which of the alternatives is true. You must not distinguish verbally hypotheses which cannot be distinguished observationally. Statements purporting to distinguish between hypotheses, etc., which it is inherently impossible to distinguish by observation are meaningless.

Example: If I assert and you deny that the far side of the moon is painted puce, we (or our descendants) can, in principle, settle the matter by taking a rocket-ship, when one is available, and going to look. But if you assert and I deny that the Absolute has relations, there is, inherently, no conceivable means of deciding between us—we cannot catch the Absolute and third-degree it as to the number of its uncles,[1]

[1] Note that this deliberately silly remark is neither more nor less sensible than the disputed proposition; but it has the advantage of being so obviously silly that no one would pay any attention, whereas many

or make any observations on 'it' at all; and to define the word 'Absolute' as referring to 'that which has no relations' would be begging the question.

4. It is futile, time-wasting, and therefore illegitimate, to talk about (alleged) entities which are, *ex hypothesi*, inherently unobservable. Such (alleged) entities should be treated as non-existent. This is not quite the same as categorically declaring that they do *not* exist; at least, not according to the ordinary usage of language, though it comes to the same thing. Actually, as we shall see in an important connexion later, it is a question of how you propose to use the word 'exist', and I think there can be little doubt that it is best to use it in such a way that we flatly deny the existence of inherently unobservable 'entities'.

For example: If I say 'I have a lump of gold in this safe as big as your head, but unfortunately it vanishes as soon as I open the door', then the position is indistinguishable from that of there being no such lump of gold in the safe; and it is simpler and safer to say firmly that the lump of gold does not exist than to aver that it does, and then be obliged to invent some elaborate rigmarole to 'account for' the fact that it can never be observed. Otherwise we shall be likely to find someone deducing the strangest conclusions from the supposed existence of the observation-defeating mechanism—as indeed has happened again and again, *mutatis mutandis*, in the history of philosophy.

In short: In the name of all sanity, and as posterity shall be our judge, let us at any cost refrain from talking nonsense.

I need hardly say that the foregoing statement of principles makes little claim to completeness, to logical precision, or to the avoidance of overlapping; but it should serve to make reasonably clear the kind of discipline to which I think we ought to submit in an inquiry of this kind.

It is perhaps worth noting at this point that these principles, though I think they would be accepted by most people as 'self-evident', are so by virtue of common sense (i.e.

people have supposed that the words 'The Absolute has no Relations' are not nonsense. They are.

condensed common experience) and not by any 'intuitive' apprehension of their *a priori* truth. Otherwise they would be 'axioms', and my adherence to them would be open to the same criticisms that I myself level against metaphysicians who profess to deduce conclusions about matters of fact from axiomatic premises. That is to say, it is a matter of common empirical observation that nothing but confusion and strife arises from the use of vague and undefined language, from one person using a word in one sense and another in another, from attempting to draw distinctions which make no difference in practice, or from acting on verbal references which have no counterpart in the world of fact. The only difference is that, in philosophical contexts, these sources of confusion are more subtle and more deeply rooted in linguistic habits.

10. *Course of the Discussion.* In the next chapter I shall go more fully into the reasons for the complete and invariable failure of metaphysicians to make progress in this (or any other) subject; and in the one which follows, I shall give a necessarily brief account of the theory of Meaning, which, as I have already indicated, is vital to the whole discussion.

This will complete the preliminary clearing of the ground, and the principal points I shall be concerned to make thereafter may be somewhat roughly summarized as follows:

1. To answer the question 'What *is* Matter?' we must examine the situation known as perceiving a material object. When we do so, we find that the only entities of the existence of which we can be absolutely sure are certain 'sensations' (e.g. visual) or 'sensa', or, to use my own term which includes images, etc., 'cognita'. We must therefore reduce everything to terms of these, and it is illegitimate to speak of matter as consisting of anything else. On any particular occasion there are associated with these cognita, by virtue of past experience, other cognita (e.g. tactile) which we 'expect' to follow them under certain conditions; and our assurance that it is a material object we are looking at depends on these 'prospective' cognita duly following on, say, certain movements (themselves expressible in terms of cognita, notably

kinaesthetic sensations) in a manner conforming to that past experience. If they do not, then the object is not material but hallucinatory, etc. That is to say, any material object *is* a cognitum sequence ('sequential pattern of cognita') of a certain type specified by the laws of physics (regularities observed by physicists), and there is no sense whatever in trying to make out that it is anything else.

2. All talk of 'things-in-themselves', 'substances', or like (alleged) entities which 'have' properties, etc., is accordingly just so much mythological foolishness (except, of course, as a convenient shorthand recognized as such) and must be eliminated from discussion. Such alleged entities are inherently unobservable, *ex hypothesi*, and must be treated as non-existent, because any observation of them would itself consist of the cognizing of cognita of some kind, which would form parts of some sort of pattern constituting an object of some kind.

3. But these conclusions are *not* equivalent to 'denying the reality of matter', *or* to making out that 'matter is really no more than mental' (or like phrase). As will have been pointed out in the chapter on Meaning, that is 'really' X which conforms to the specification of X; and that is 'really' matter which conforms to the specification of matter as laid down by physicists. Matter is as matter does.

Unfortunately, physicists are, as a rule, the most inveterate of mythologists; for they commonly persist in maintaining that they are studying certain 'substances' which 'have properties', whereas, like everyone else in every field, they are actually studying cognitum sequences of certain types.

4. In particular, physiological psychologists begin at the wrong end. What is 'given' is *not* first receptors, then nerves, then brain cells, and finally 'sensations', 'sensa', or 'consciousness', *but* cognized cognizables, *alias* cognita.

5. A mental object *is* a cognitum sequence of the type specified by the laws of psychology, so far as these are known. Inasmuch as our present knowledge is very limited, there is no objection to saying that it *may* be found convenient, or even necessary, to subdivide cognitum sequences of

non-physical type into phenomena of, say, 'spirit' as well as of 'mind'. Every phenomenon must reduce to cognitum sequences of some kind, regardless of how we label it: there is nothing else for it to reduce to.

6. Just as it is absurd to talk about inherently unobservable 'things' which 'have' properties, etc., so it is meaningless to talk about 'minds' or 'egos' over and above the cognita or cognizables that would usually be said to form their 'content'; these must equally be eliminated from fundamental discussion. What we call 'a mind' *is* all those cognita or cognizables which would ordinarily be said to constitute the content of that mind, i.e. cognita in the case of what we call 'conscious states', cognizables in that of 'the subconscious', etc. There is nothing to prevent a cognizable or cognitum being a constituent of two or more minds.

7. It follows that 'consciousness' cannot be a 'property of the mind', or 'being conscious of' or cognizing an activity of it. Consciousness *is* that state of 'tension' (the word is used analogically) between the cognita constituting the object, etc., which the mind would ordinarily be said to be 'conscious of', and those which would ordinarily be said to form the 'content of' that mind.

8. A cognitum is not only a part of the entity cognized, it is also a part of the cognizing mind.

9. The difficulty of how Mind can act on Matter or vice versa, therefore disappears. Matter and Mind are not qualitatively different 'orders of reality', or the like, but inseparably interlocking and interwoven organizations or patterns of the neutral and irreducible entities or 'particulars' (cognizables) of which the universe is composed.

10. The essential difference between philosophic materialism and philosophic idealism is purely linguistic, and depends on the way in which we decide to use the word 'exist'. The basic requirements of the materialist (illegitimate dogmatism apart) are sufficiently met by supposing (as is also most economical) that these 'particulars' exist as 'cognizables' when not actually cognized as 'cognita'.

This schedule of points that I hope to make again does not

profess to be perfect in form, substance, or comprehensiveness; but it should be sufficient to indicate the kind of lines on which I propose to work, and the kind of conclusions at which I shall arrive.

I shall also devote a certain amount of attention to the recently investigated phenomena of Psychokinesis, which appear to show that what may reasonably be termed the direct action of mind on matter is a fact in nature under suitable conditions. And I shall attempt some discussion of the way in which mental phenomena in general and those of what we call an individual mind in particular seem most likely to be regulated. But much of this will be frankly speculative.

11. *Difficulties.* I should be sorry if any reader were to conclude from the foregoing that I am proposing to deal in peculiarly abstruse or difficult conceptions. This is not the case. There is nothing very abstruse in refusing to accept the existence of or to argue about the nature of something which can never be observed; and so far as there is any difficulty at all, as regards fundamental principles at least, it is in ourselves and not in the subject-matter.

There is no heavy spade-work involved here, as there is in studying the Russian language, say, or the theorems of spherical trigonometry; but none the less a considerable effort is likely to be called for, notably in the matter of facing and dealing with internal resistances arising from traditional habits of thought and linguistic conventions. The difficulty, I think, is of much the same kind as that involved in realizing that the earth is not a flat plain but a sphere—with the antipodeans 'hanging head downwards'—and does not rest on anything but floats unsupported in space. There is nothing inherently difficult about these conceptions, in the sense that they are intelligible as soon as stated, but many people have found it far from easy to assimilate them.

There should be no more difficulty in abstaining from belief in unobservable 'things', which 'have properties', and the rest of the metaphysical mythology, than in not believing that a mischievous imp animates the soap that eludes us in our bath-tub. But primitive man took it for granted that every material object had an animating 'soul', and it has taken us many generations to free ourselves from such superstitions —even now not so completely as we might—and somewhat similarly as regards our philosophic thought we remain lost in the maze in which the conventions of language have imprisoned us.

To escape from the maze, into the free air where clear thinking is possible, it is not necessary to thread the devious passages of an intricate labyrinth; it is rather a matter of taking our thought, as it were a horse between our knees, and lifting it straight and clean over the confining fences. And this is not a question of subtlety and cleverness so much as of resolution and fortitude.

To change the metaphor: in this matter we are, it seems to me, very much in the position of a victim of persecution mania. The world, he is sure, is full of enemies determined to beset him with every imaginable difficulty. It is true that he cannot see them, or find the slightest rational evidence of their existence; but this only proves their extraordinary cunning, so that, to keep his suspicions plausible, he is forced to devise ever more and more correspondingly complicated theories to explain away the lack of evidence. Could he but bring himself to make the effort, he would enormously simplify his life by realizing that the plain and obvious reason why he can never locate his enemies, or find proof of their machinations, is not that they are inconceivably subtle and the problems they set him difficult beyond his powers, but that there are no enemies and no problems to solve. And for the wretched metaphysician, harried by phantasmal conundrums of his own devising, the one all-embracing solution is to wake up and find them all a dream.

II

THE FAILURE OF METAPHYSICS

In the beginning was the Word . . .

The Gospel according to St. John, I, 1

. . . it is a tale
Told by an idiot . . .
Signifying nothing.

SHAKESPEARE, *Macbeth*, V, 3

There is nonsense, damned nonsense, and Metaphysics. W. W. C.

12. *The Fact of Failure.* If anyone doubts that metaphysicians have completely failed to throw even a glimmer of light upon the alleged problems with which they purport to deal, let him go out and try to find a text-book of Metaphysics. He will find plenty of excellent text-books, in any reasonable language, on Electricity, Astronomy, Physiology, Chemistry, Hydrodynamics, and every scientific and technical subject under the sun; but not on Metaphysics. Or, if he happens to come across one so entitled, he will soon discover that its contents are very different from those of text-books on other subjects.

A comprehensive treatise on Chemistry, for example, apart from being crammed with facts, which his book on Metaphysics is not, is likely to begin with a historical survey of the subject. It will trace developments from, say, Thales, who declared that 'all is Water', through Aristotle and other exponents of the Four Elements Theory (Earth, Air, Fire, and Water) to the medieval Alchemists; and thence by way of the Phlogiston theory, to the experiments and conclusions of Lavoisier, Black, Priestley, and Dalton and the foundations of the modern science. And such a survey will presumably end with some account of the latest views on the constitution of atoms—protons, electrons, neutrons, quantum theory, wave-mechanics, and the rest of it—and the artificial transmutation and generation of elements. The reader will be shown the evidence on which the views of different periods were based, and why some of them had to be abandoned or modified. He will be told what is, at the present time, regarded as certain, what is still in doubt, and what problems are still to be solved. Almost from start to finish the story will be punctuated by accounts of positive discoveries, and achievements in analysis or synthesis; and it will be made clear how the whole science has been built up by experiment and reasoning—'precept upon precept, line upon line'—through the contributions of innumerable workers. And even if he consults a treatise on Mathematics, or any branch thereof, he will find very much the same thing; only here it will not be a record of experiments, but of formal deductions from definitions. There will, however, be essentially the same story of discovery and development and of contributions from many sources.

But the earnest inquirer will not find anything of this sort in his alleged 'text-book' of Metaphysics (or of Philosophy either, apart from formal Logic). He will find nothing but a record of what various people *said*; and he will note that, although there is a certain sameness (mainly a matter of unintelligibility) about their sayings, no two of them agree. He will search in vain from A to Z—or at any rate from Zeno to Samuel Alexander—for any report of discovery or of experimental proof. Nowhere will he find such sentences as 'So and-so *found* that . . .', 'someone else *demonstrated* . . .', 'such-and-such a theory had to be discarded, *because* . . .'

Unless he is an inveterate wish-thinker and self-deceiver, or imbued with a more than dog-like faith in the infallibility of philosophers, he will come to the obvious if distressing conclusion that the reason for these omissions is simply that there is *no* positive achievement to be reported, no demonstrations to be put on record, and no progress to be announced.

Having paid his money, he may then take his choice between three (and I think only three) possible explanations;

first, that the problems of metaphysics are so difficult (in the same sort of sense that the problems of physics, chemistry, or physiology may be difficult) as to defy the most brilliant human intellects for more than two thousand years; second, that philosophers have, for one reason or another, invariably tackled them the wrong way; third, that they are not, properly speaking, problems at all. But that the metaphysicians have signally failed in their self-appointed task of elucidating the Ultimate Nature of Reality, the Final Causes of Things, and so forth, can hardly remain in doubt.

13. *The Reasons for Failure* (1); *Axioms.* I have already given one very good and amply sufficient reason why metaphysicians have never achieved any positive result in their own field, namely, that any worthwhile conclusion they may, by accident or otherwise, arrive at, is automatically subsumed under one or other of the special sciences if it is about an observable matter of fact, or under some branch of Logic or Mathematics if it is about the use of symbols. But I think it is important to examine what has been happening rather more closely than this. The great metaphysical philosophers were about as far from being mentally defective as any men who ever lived; and it is not at all easy to see on first inspection why men of such exceptional intellectual prowess should have persisted as they did (and still do), generation after generation, in a quest which seems to many of us to-day as inherently hopeless as the attempt to reach the rainbow's end. As might be expected, the answer is to be found in a mixture of history and psychology.

The whole development of our intellectual thought has been moulded by the influence of the ancient Greeks, just as our religious thinking has been by that of the ancient Jews. But, outside of art, literature, and politics—that is to say, in all its more formal aspects—Greek thought may fairly be said to have been 'intoxicated with the exuberance of its own geometry'. To this science the Greeks, as we all know, made the most brilliant contributions, and it is only natural that geometrical reasoning should have been regarded by them as the very ideal and pattern of all right thinking—'Let no one ignorant of geometry enter my doors',[1] and so forth. And probably this influence was strengthened rather than weakened with the passage of time. There are fashions in art, literary tastes may change, political ideals vary from people to people and from climate to climate; but the square on the hypoteneuse remains for ever equal to the sum of the squares on the other two sides.

Now the Greeks believed, almost inevitably, though not quite correctly, that the methods of geometry had enabled them to discover facts of importance about the physical world in which they lived. What more natural than that they, and those who inherited their learning, should conclude that the same basic methods as had led them to these beautiful and often useful results would be equally efficacious in other fields of inquiry—nay, that they alone could be so? Thus it is in no way surprising that philosophers, seeking to penetrate what they felt were the profoundest mysteries of the universe, should follow as closely as they could the brilliantly successful method of Greek geometry.

What is this method? Essentially, it is simply this: you start by laying down certain propositions (axioms or definitions) of which the truth appears to you to be unchallengeable: 'A straight line is the shortest distance between two points'; 'Two straight lines cannot enclose an area'; 'A circle is the curve traced out by a point moving so that its distance from a given fixed point is constant'; and so forth. From these, by strictly logical processes and not using anything not implicit in them, you deduce certain conclusions—as that the angle in a semicircle is a right angle. You regard these conclusions with absolute assurance; for your logic is impeccable—this is usually easy enough to check—and your premises irrefutable.

All metaphysicians have followed this procedure of arguing to conclusions, purporting to be relevant to the world of fact —as that 'God exists', or 'the Soul is immortal'—from axioms and definitions which they claim to be indisputably true on grounds other than those of empirical observation.

[1] There is a tradition that these words were written over the door of Plato's Academy. (H. H. P.)

But can any such axioms, etc., safely be depended on? Are any 'truths' properly to be called 'self-evident', other than those purely verbal tautologies—'A is A'; 'A is not not-A', etc.—to deny which is to state a formal contradiction in terms? Surely there are, it will be urged; can it, for example, be anything *but* self-evident that two straight lines cannot enclose an area? The answer is: Only if the term 'straight line' be so defined as to make it so; i.e. only if you define a straight line as a line of such a kind that an area cannot be enclosed by two of them—or disguisedly to the same effect—not if you define it as 'the shortest distance between two points'. This is obvious if you are dealing with the surface of a sphere and the rules of the game require that all your lines lie wholly in that surface (short cuts through the body of the sphere being barred). In this case the shortest distance (over the surface) between any two points is an arc of the great circle[1] passing through them; and two lines defined by this shortest-distance criterion can perfectly well enclose an area, as do the meridians of longitude on the terrestial globe.

The illustration is elementary and may not appeal to purists; but it should serve to illustrate the essential point, namely that, in geometrical reasoning, the 'self-evidently' true axioms are, in fact, reliable only *if* the space you are interested in happens to be of the kind you have assumed. This comes to the same thing as saying that the space for which the conclusions are true is *defined* by the axioms with which you start: any set of (geometrical) axioms which is not inherently self-contradictory will serve to define a space of some sort.[2]

Now the axioms used in Greek (Euclidean) geometry are not self-contradictory, and they accordingly define a certain kind of space. For this space the conclusions deduced are absolutely true, by definition; but not (or not necessarily) for any other sort of space. As it happens, the physical space in

[1] A 'great circle' on a sphere is any which passes two points of which one is diametrically opposite the other.

[2] Cf. Bertrand Russell, *Introduction to Mathematical Philosophy* (Allen and Unwin, London), pp. 38 ff.

which we live is very nearly of the kind defined by the axioms, so that no ordinary process of terrestial measurement will show any discrepancy. Consequently, both the conclusions reached and the axioms on which they depended were not called in question for a very long time; and the axioms were apparently so much a matter of common experience that no one hesitated to accept them as 'self-evident', though actually they were not. It was not till some two thousand years later that geometricians (notably Riemann) showed that other sorts of space, with corresponding geometries, were theoretically possible, and that astronomical measurement (especially in connexion with Einstein's theory of Relativity) empirically verified the hypothesis that the actual space of the physical world is not strictly Euclidean in its properties.

Probably no heavier blow has ever been struck at metaphysics, though I do not suppose that anyone realized it at the time (and not many now); for it at once became clear that conclusions deduced by even the most rigorous logic from even the most (apparently) 'self-evident' axioms were not necessarily applicable to the world of observable fact. It is true that in the particular case of Euclidean geometry the discrepancies were negligible for all terrestial purposes, but the inherent untrustworthiness of the whole procedure as a guide to the real world was none the less clearly demonstrated.

14. *Unreliability of Axioms, contd.* The point is so important that I must be forgiven if I approach it again from a different angle, mainly for the benefit of the wholly unmathematical reader, who may have been alarmed by all this talk of geometry and different sorts of space.

The metaphysician wishes to arrive at some reliable conclusion (or more usually, if unwittingly, to 'prove' some conclusion he has already adopted) about the 'nature of Reality', or 'God', or the 'Soul', or some such words. But even a metaphysician cannot argue completely *ex nihilo*: he must start somewhere and from some sort of assumptions. Naturally, he is anxious to choose assumptions which he is confident no one can overset, or even be likely to dispute; for he knows very well that, if he gives them a chance, his brother metaphysicians will be on to him like a pack of wolves, and rend him and his cherished arguments limb from limb. He therefore tries to confine himself to assumptions (axioms, definitions, etc.) which are 'self-evidently' true.

Unfortunately, no proposition is self-evidently true except in so far as it is purely tautological, i.e. a mere rearrangement of symbols or a substitution in which one symbol is replaced by another precisely equivalent to it. If it is anything more than this, then, tacitly or explicitly, and whether the metaphysician realizes it or not, it must contain some kind of statement about the factual world; and this will need verification to make sure whether it is correct. In the absence of verification (or relevant evidence) we have no means of corresponding doubt as to the validity of any conclusions deduced from it.

Such factual implications may arise in very insidious ways. Spinoza, for example, takes as his first axiom the proposition 'All things which are, are either in themselves or in other things'.[1] On the face of it this is a purely tautological (analytic) proposition of the form 'Every A is either B or not-B', which no one could dispute—and incidentally, as already insisted, could lead to no conclusion about the factual world; but it clearly implies that every 'thing' must be 'in' something, viz. either itself or something else. Now I do not in the least know what Spinoza meant by saying that a thing is 'in' anything at all, and venture to doubt whether he did; but the proposition is unquestionably implicit in the axiom, and is certainly a statement about a matter of fact (if it means anything at all) and not about symbols. As such it may be true or false (assuming it to have meaning); but we cannot possibly tell which without some process of factual observation.

Rather more obviously, Spinoza's definition of *Substance* as '. . . that which is in itself and is conceived through itself'[2] tacitly implies, the moment it is used as an aid to arriving at factual conclusions, that there does exist at least one entity in

[1] Spinoza, *Ethics* (Everyman's Edition, reprint of 1941), p. 2, Pt. I, Axiom I.

[2] *Ethics*, Pt. I, Definition III.

the universe which answers to the definition; and this again can only be assured by observation.

More generally, to lay down any definition intended for use in a deductive argument leading to a conclusion about matters of fact, other than the definition of a symbol (word, etc.) in terms of other symbols, is to imply the factual existence of the entity so defined (otherwise the definition is futile), and this constitutes a statement of fact about the universe, which is not necessarily true but requires verification. Definitions of symbols in terms of other symbols are only relevant to factual discussions if these other symbols themselves refer to existent entities, which leads us again to the same need for verification.

It follows, I think, that no reliable conclusion about matters of fact can possibly be deduced from axioms, though such conclusions *may* be true *if* all the factual propositions implicit in the axioms are of a verifiable character. Thus the metaphysician, did he but know it, is in the same position as the mathematician in Russell's famous epigram, who 'never knows what he is talking about, or whether what he is saying is true'.[1] Properly speaking, he should never go further than saying, 'IF there be existents in the universe which conform to the definitions I am using, or of the kind implied by my axioms, *then* such and such conclusions are true of those existents'. Unhappily, like Eddington's 'super-mathematician', he does not even know what he is doing: he fondly imagines that he is talking about existent entities, whereas all he is actually doing is to shift words about into different patterns on the metaphysical chess-board. The entities to which these words are supposed to refer may or may not exist in the universe. If so, well and good; but, if not, then (as one might say) *Ex verbis nil nisi verba.*

15. *Reasons for Failure* (2); *Words.* I suppose that the four characteristics that have chiefly enabled man to achieve his dominant place in the world are the erect posture, the opposed thumb, the hypertrophy of grey matter at the top of the spine, and the power of speech (including writing);

[1] *Mysticism and Logic* (Longmans, London, 1921), p. 75. Note that this statement is not metaphorical or analogical, but strictly accurate.

and I should feel inclined to surmise that of these the last is by far the most important (except in so far as the third may be prerequisite to it). That is to say, I should think that almost any of the higher animals, given the power of speech, would have acquired an almost equal ascendancy, in the absence of man himself, even if they lacked the first two advantages.

The reason is that it is by the use of speech, and writing at a later stage, and by these alone, that he has been able to transmit the accumulated experience of the species (or rather a selected and concentrated distillate thereof) simply and compendiously from one generation to another, instead of being obliged to wait, so to say, for the slow processes of natural selection and any others there may be, to weave a fraction of it into his make-up in the form of instincts or inherited reflexes. By the aid of speech and writing every normal person has an immense amount of vicariously acquired knowledge at his disposal, so that it is not necessary for him, as with other animals, to learn practically everything from the ground up.[1]

It is reasonable to suppose that even primitive man realized the advantages accruing from the use of language, in some degree, and was thus at least predisposed to accord a certain measure of veneration to this almost miraculous gift of verbal communication; but it seems to me probable that other and more important factors also operated to raise the prestige of *words* to the status of quasi-divinity.

There is considerable reason for supposing that primitive people, and at least some young children even to-day, not infrequently have genuine difficulty in distinguishing between visual images (of memory, hallucination, or imagination proper) and the sight of material objects; and it has even been suggested[2] that this realistically vivid imagery was the original and natural type. If so, it must have had very serious drawbacks, to the point of possessing negative survival value. It

[1] For emphasis on this point, and fuller treatment, see Korzybsky, *Science and Sanity* (New York, 1933).

[2] Cf. E. R. Jaensch, *Eidetic Imagery* (Kegan Paul, London, 1930), pp. 21 ff.

would be trying enough laboriously to stalk an imaginary aurochs in the belief that it was a material one; but the reverse error in the case of a sabre-toothed tiger might well be fatal. On a smaller scale, a correspondent tells me that as a child she was often scolded for talking about people whom she declared she could see but, who, to the eyes of her elders, were not 'really there' at all. The tendency may accordingly be supposed to have been suppressed at a relatively early stage; but it may well be largely responsible for the comparatively extensive 'seeing' (and even hearing, etc.) of ghosts, demons, nature spirits, and the like in unsophisticated societies.

Bearing this in mind, note next that to hear the name of a familiar object (or class of objects) does, in fact, tend to call up an image of that object (or of a more or less typical member of that class). It is true that a modern educated man, accustomed to use extensively a highly developed language, does most of his thinking in words, or images of them, rather than in images of concrete objects. In such a case the word 'cat', for example, probably calls up mainly verbal forms (sentences, etc.) having to do with cats, rather than a definite image of a furry, mewing, etc., quadruped, though images of a particular or of a typical animal, or of sensations of softness, warmth, furriness, etc., will not be far away. But I should think it almost certain that in primitive and barely articulate man, for whom the enunciation of a single sentence, or even a single word, must have been quite an undertaking, the revival of visual, etc., images of the concrete object will have been the almost exclusively predominant effect.

Under such conditions, or approximately so, where the hearing or enunciation of the word 'aurochs' (or whatever uncouth eructation served the purpose in those days) instantly evoked an image of the animal—an image perhaps so vivid, as I have suggested, as to be hardly distinguishable from the actual sight of the creature—the *word* and the *thing* must have appeared inseparable to the very point of identity. To most intents and purposes the word *was* the thing, or at least a part of the thing; for there is no reason to suppose that primitive man was capable of any considerable feats of analysis, and even to-day identification of word and thing is prevalent to a serious if usually unrealized extent.[1]

It is interesting to note how elegantly the foregoing explains the still lingering belief that knowledge of a person's *name* gives you power over him. When you pronounced (or thought of) the *name*, an image of the person appeared; this was so vivid as to be almost as good as the presence of the person himself; in effect, pronouncing the name summoned the person; that is to say, you had caused him to obey you—you had power over him, q.e.d.

Now imagine this kind of thing going on, approximately as described, with language consisting in the early stages, one may suppose, of little but the names of material objects and (as we say) their properties and of observable activities. At this level there will always *be* a 'thing', i.e. some sort of an observable object—or, in the case of words like 'on', 'inside', 'behind', etc., some sort of observable and characteristic configuration of objects—to correspond to any given word. More technically, there will be no difficulty in identifying to what the word refers. Hence will arise a conviction, probably never expressly formulated, but none the less strong for that, based on an immense amount of empirical evidence, that for any word there *must* be a corresponding existent (object), even if, for one reason or another, one cannot actually see or touch it at the moment—and, in any event, the word evokes an image which does very nearly as well. And the same conviction will tend to persist when it is found convenient to introduce what we call 'abstract' words, such as 'courage', 'loyalty', treachery', etc., or their humbler primitive forerunners. These are words like the others, and they, too summon up images of sorts, though doubtless vaguer and more mutable than those evoked by the names of material objects, etc., and for them too, or so primitive man will argue or take for granted, there must be identifiable existents in the factual world.

[1] Again see Korzybsky, loc. cit.

Now sophistication does not necessarily bring wisdom or critical power, and the forces of superstition and the traditions of magic die hard; indeed, except they be most resolutely challenged, they never die at all. Thus the naïve belief (or rather 'taking for granted') that for every word used in a language there must be a corresponding and identifiable existent in the factual universe is only too likely to persist unless active steps are taken to counteract it; and metaphysicians have only too easily fallen into the error of supposing that, because they can find such words as 'substance', 'absolute', 'reality', 'essence', and the rest in their dictionaries, they must be talking about *something* when they use the words in discussion. This supposition is comprehensible enough and even pardonable, in view of what I have just said; but it is a complete fallacy, and unless the existents allegedly referred by the words can, in principle at least, actually be identified by some sort of observation, they are, in fact, talking about nothing at all.

16. *Reasons for Failure* (3); *Grammar*. A language does not consist of words alone; there must also be rules for modifying and arranging them (grammar and syntax), or otherwise so contriving that the sentences in which they occur correspond correctly to the relations between the objects, etc., referred to. Thus the sentences 'Dog bites man' and 'Man bites dog' refer to two importantly different situations, as every journalist is taught in his cradle. In English this difference is expressed by changing the order of the words; but in Latin, and other more highly inflected languages, one or more of the words themselves is altered, and variations in order become available for more subtle purposes of emphasis, etc., without relevance to the situation described. The words *canis*, *mordet*, and *hominem* will always mean (i.e. be translated as) 'Dog bites man', whatever the order in which they are arranged, and never 'Man bites dog', for which we need *mordet*, *canem*, and *homo*, with equal indifference to order.

Now, just as metaphysicians and others have taken for granted that to every word in the language there must correspond some existent in the world of observables, so there is a tendency to take for granted that the *structure* of the factual world must correspond to the structure of the language they use in talking about it, i.e. of the sentences purporting to describe it.[1] In fact, I am inclined to think that this second tendency is, and naturally would be, even stronger than the first. Anyone can see that it would be possible to invent a word (e.g. 'Jabberwock') having no corresponding factual referent, and that, in suitable circumstances, it would be possible to induce a widespread belief that it had (as in propaganda), or that such a belief might arise naturally, through wishful thinking and like causes. And no great acuity is required to realize that it is logically pernicious, and no more relevant to serious discussion than mere punning, to try to base conclusions on the accidental or merely historical form of words—as Chesterton did, in one of his less responsible moments, when he wrote about the 'weak and inconclusive' word 'compromise' containing the 'strong and ringing' word 'promise'.

But the correspondence of factual structure with grammatical structure is more subtle. There is no doubt about it in the case of such a sentence as 'Jones bites biscuit'. Interpreted by the accepted rules of English grammar, it implies that there is an identifiable existent referred to by the symbol 'Jones', another, quite distinct from this, referred to by the symbol 'biscuit', and an asymmetrical relation between them, to wit a certain unambiguously identifiable activity, referred to by 'bites'; and the order of the words tells us which end of the relation, so to put it, applies to which *relatum*. Similarly, the sentence 'Jones has a watch' refers to a situation in which there is a different kind of asymmetrical relation between the two equally identifiable and separable referents of 'Jones' and 'watch'. There may possibly be some ambiguity about the precise meaning of 'has', for the watch may be, as we say, Jones's legal property, and it may be on his wrist, or left behind in his house, or locked up in his bank; or he may have just picked somebody's pocket. In any

[1] Cf. e.g. Russell, *Our Knowledge of the External World*, first edition (Open Court Co., Chicago and London, 1914), pp. 45 ff.

of these cases we might well use the word 'has' to refer to the relation between Jones and the watch; but whatever doubt there may be here, there is none about the distinctness of the referents related, or the propriety of talking about a relation between them.

But difficulties begin to arise with such a sentence as 'Jones has a stiff leg', or 'Jones has a vile temper'. So far as *grammar* is concerned, these sentences imply that exactly the same kind of 'possessive' relation holds between Jones and his leg, or Jones and his temper, as between Jones and the watch, and that the referents of the symbols are equally clearly separable. There is nothing whatever in their structure to suggest that the factual set-up, if I may use the term, is in any way different in the three cases. But it is. The watch is no *part* of Jones—at least, not by any reasonable standards; but the stiff leg, or the vile temper *is*. Any reasonably complete specification of Jones would include a mention of his leg or his temper, and if either were eliminated he would not be, as we say, the same man as before; and if all such components, similarly related,[1] were eliminated, there would be precious little Jones left. The referent of the symbol 'Jones' is *not* sharply separable in the world of fact from the referents of the symbols 'stiff leg' and 'vile temper' in the same way as it is from that of 'watch'; and it is simply not true to say that it is, as is implied by the structural identity of the three sentences considered. But an intelligent Martian, confronted with the three sentences 'Jones has watch', 'Jones has leg', 'Jones has temper', but lacking an English-Martian dictionary, would be bound to conclude, from a study of the grammatical structure alone, that these three sentences described or referred to three situations as identical in the relations between their constituents as would the three sentences 'Jones has watch', 'Jones has pen', 'Jones has key'. And this would not be true.

[1] Somewhat anticipating the course of the discussion, and to meet possible criticism, it should be noted that the relations are strictly between the 'components' or 'attributes' themselves, not between each of them *seriatim* and some kind of a mythological unobservable known as Jones.

17. *Grammar, contd.* Let me develop this point a little farther. The rules of English grammar are such that the sentences 'Jones has a vile temper', Jones is vile-tempered', 'Jones has the attribute (or quality or property, etc.) of being vile-tempered', 'Jones has the attribute (etc.) of vile-temperedness' are identically equivalent. Each may be substituted for the others without any change of sense; that is to say, no more and no less and no different information is conveyed by one than by another, and which we choose to use is purely a matter of euphony, convenience, style, etc.

Similarly the sentences 'The tomato has a red colour', 'The tomato is red' (or 'red-coloured', if you prefer to keep the parallelism exact), 'The tomato has the attribute (etc.) of being red', 'The tomato has the attribute (etc.) of redness' are interchangeable without alteration of sense. We might equally well say '(The) tomato has redness' without adding to or subtracting from the information conveyed, just as we say 'Jones has (a) watch'; or, conversely, 'Jones has the attribute of watch-owningness.' However we juggle the identically equivalent sentences about in the two cases, we can always find a pair that match. Hence, the Martian student would inevitably conclude that the factual set-up, or relational structure, referred by 'Jones has a watch' is identically similar with that referred to by 'The tomato is red', for they are perfectly interchangeable with 'Jones is watch-owning' and 'The tomato is red-coloured', respectively. He could do nothing but infer that Jones is to Watch as Tomato is to Redness (or equivalent phrase), and he would say that in each case there are two distinct entities—Jones, Watch; Tomato, Redness—between which the same kind of relation holds.

But this conclusion, though a necessary inference from the study of the grammatical structure, on the assumption that there is a one-to-one correspondence between the grammatical and factual structures, is false, and therefore the assumption is erroneous. For the redness of the tomato (and, of course, its roundness, squashiness, pippiness, etc., also) is not separable from the tomato in the same way that the watch is separable from Jones. You could strip Jones of all his possessions and still have Jones left (though perhaps more vile-tempered than ever); but if you take away all the attributes from the tomato, what remains? And echo answers 'What?'

These childish examples should serve to show, I think, how very dangerous it would be to rely on grammatical forms as a guide to the Nature of Reality, or whatever it is that metaphysicians are seeking to discover. Yet on the basis of scarcely less blatant puerilities have been largely erected some of the most theoretically formidable and practically influential philosophical systems that the world has ever known.

The Fascist-Nazi conception of the State as a kind of monstrous demiurge overriding, body and soul, the individual human being, together with the accompanying view that Might is logically and philosophically Right, is largely derived (I understand) from the doctrines of Kant via Fichte; and the same is even more plainly true of the system known as 'dialectical materialism'—developing from Hegel through Feuerbach, Marx, Engels, and Lenin—which is the official philosophic creed of Russian Communism.

In this connexion the following passage from Bertrand Russell[1] is illuminating. After pointing out that what Hegel calls 'logic' is 'an investigation of the nature of the universe, in so far as this can be inferred merely from the principle that the universe must be self-consistent', Russell says later 'Hegel's argument in this portion of his *Logic* depends throughout upon confusing the "is" of predication, as in "Socrates is mortal", with the "is" of identity, as in "Socrates is the philosopher who drank the hemlock". Owing to this confusion he thinks that "Socrates" and "mortal" must be identical. Seeing that they are different, he does not infer, as others would, that there is a mistake somewhere, but that they exhibit "identity in difference". Again, Socrates is particular, "mortal" is universal. Therefore, he says, since Socrates is mortal, it follows that the particular is the universal—taking the "is" to be throughout expressive of identity. But

[1] *Our Knowledge of the External World*, first edition, p. 38.

to say "the particular is the universal" is self-contradictory. Again Hegel does not suspect a mistake, but proceeds to synthesize particular and universal in the individual, or concrete universal. This is an example of how, for want of care at the start, vast and imposing systems of philosophy are built upon stupid and trivial confusions, which, but for the almost incredible fact that they are unintentional, one would be tempted to characterize as puns.'[1]

This is not so very different from our story about Jones and the tomato. If Hegel had studied the matter he would doubtless have come portentously to the conclusion that, because the sentence 'Jones had a watch, but he pawned it' may refer to an identifiable occurrence, the sentence 'The tomato had a red colour, but it is at present in the hock-shop' is not nonsense. And after two thousand years of philosophizing, we are expected to pay serious attention to the theories of people who run around chattering about 'the interpenetration of opposites', and like slogans of Hegelian provenance.

Properly speaking, it is no more illogical, though it may not sound so silly, to say that a tomato may pawn its colour, than it is to infer from the form of the sentences 'The tomato has a red colour', etc., that there is a 'something', existing in its own right, so to say, and to be known as 'the tomato' (or 'the substance of the tomato', or 'the tomato itself', etc.) distinct from but somehow possessing the 'redness'; and the fact that this is standard practice is merely so much the more unfortunate.

I shall have much more to say about this in Chapter IV below, and in Chapter VI shall insist that the same considerations apply to the supposed Self, Ego, Soul, Mind-itself, etc., misconceived as existing apart from, and in some sense 'possessing' what grammatical conventions cause us to

[1] *Our Knowledge of the External World*, first edition, p. 39 n. Note that if the metaphysician does not seek to arrive at conclusions about matters of fact, about the universe, etc., he is not a metaphysician, but a pure logician or mathematician; and if he does not rely on axioms, etc., claimed to be true *a priori*, but on premises derived from observation, he is again not a metaphysician but a natural scientist.

describe as 'its' contents, experiences, states, thoughts, feelings, or whatnot.

18. *Summary*. Meanwhile, I think we may sum up as follows. Metaphysics may be defined as the attempt to reach conclusions about matters of fact (notably the nature, structure, etc., of the universe as a whole) from *a priori* premises; that is to say, from axioms, definitions, etc. assumed to be indisputable and self-evidently true. Any such attempt is foredoomed to failure, because the only propositions which can be known to be true independently of experience are 'analytic' (tautological) propositions dealing only with the use of symbols, of which the truth is assured by the definitions of the symbols themselves (i.e. they are true only because we have taken steps in advance, in the process of definition, which automatically ensure that they shall be), and these can have no relevance to matters of fact. Axioms not falling in this class, however they may be disguised, are necessarily 'synthetic'; that is to say, they are themselves concerned with matters of fact and accordingly require empirical substantiation.

When we ask *why* so many men of outstanding intelligence and ability have none the less failed to realize this, and have undertaken these inevitably fruitless quests, the answer is to be found mainly in the linguistic factors I have been discussing.

About the last thing a bird will discover, one may suppose, is the density of the air in which he flies—indeed it was only demonstrated by Torricelli in 1643—yet by virtue of it he may be carried far off his course. Somewhat analogously, the vocabulary and grammar of the language he uses may exert, as it were, a great though unrealized pressure on the metaphysician and lead him very far astray. So long as he omits to study the mechanism of language—and particularly the theory of meaning—he has no more chance of finding out what is happening to him than a drifting balloonist who can see neither the earth nor the stars.

In these circumstances it is no use studying the works of even the most famous metaphysicians, in the hope of discovering that some are right and others wrong. None of them are either right or wrong; the conclusions of all alike are necessarily meaningless and void. With the greatest admiration for the integrity, ability, and learning of the great philosophers, and with the fullest appreciation of their logical, humanitarian, aesthetic, hortatory, stimulative, etc., contributions to mankind, we must respectfully pass their metaphysical effusions by on the other side, and see what we can do with a fresh start and a different line of approach.

III
MEANING

It all depends on what you mean by . . .

C. E. M. JOAD, in B.B.C. *Brains Trust*

'The question is,' said Alice, 'whether you *can* make words mean different things.'

'The question is,' said Humpty Dumpty, 'which is to be master—that's all.'

LEWIS CARROLL, *Through the Looking Glass*

19. *General.* Opinions may differ as to the precise place in the philosophic Pantheon that posterity will allot to Professor Joad, but there can be no doubt that he has rendered a great public service in impressing on many thousands of people the fact that the meaning of a word is not necessarily fixed and unalterable, but may vary from one user to another; and that this may have an important effect on the outcome of debate.

Innumerable discussions fade away into inconclusiveness, or break up in acrimony, simply because the participants, though using the same words, are in fact talking about different things, and are accordingly foredoomed to misunderstanding and cross-purposes; that is to say, because they are not agreed on the meanings of the words they use. Politics—democracy (what do you *mean* by 'democracy'?)—freedom (what do you mean by 'freedom'?)—freedom is good (what do you mean by 'good'?), and so forth interminably.

Very shortly we ourselves will be at grips with our subject, discussing 'Matter' and 'Mind', and if we are not crystal clear as to what we mean by these words, all these preliminaries will have been wasted. But we cannot take the necessary steps to find out what these words mean (i.e. what we are talking about) without at least some discussion of the whole question of Meaning.

Let me try to make the reason for this clear, so far as is possible at the present stage. To say that it is important to know what we are talking about, or what we mean by the words we use, is only common sense; but it is one thing to agree on this and quite another to find out what we really *are* talking about; indeed this is the core of the whole business. Evidently, the whole question turns on what Matter and Mind actually *are*; and since this discussion, like all others, is necessarily conducted in words, this is equivalent to ascertaining precisely what it is to which we are actually and irreducibly referring when subject to the accepted conventions of the language, we use the word 'Matter' or the word 'Mind'. Common sense would suggest that we all know what these words mean; and that, if we do not, we should consult a good dictionary.[1] But if we do we find: *Matter* = 'Substance(s) of which a physical thing is made'; *Mind* = 'Seat of consciousness, thought, volition, and feeling' (*Oxford Concise*), or worse. I defy anyone to reach any conclusion regarding the relation between Mind and Matter from such a basis as this (or from the definitions in any existing dictionary), while *Webster's* flat statement that 'The nature of matter is unknown . . .' is even more discouraging. Even if we consult a physicist, whose trade it is to study matter, we shall find that, although he can tell us a great deal about the *properties of* Matter, he will not even try to tell us what matter *is*, unless he has travelled much farther along the path of the present discussion than is usually the case; and the same is true, only more so, if we ask a psychologist what mind *is*.

It is clear that we must go far beyond the dictionary if we are to ascertain what we are trying to talk about—i.e. what the words 'matter' and 'mind' actually do mean—that is to say, we must try to find out how it is that words can properly be said to 'mean' anything at all, why different words 'mean' different things, and so forth—in short, we must study the 'meaning of *meaning*'.

[1] This, in effect, is what the metaphysician does, but with the added complication that he usually relies on a private dictionary of his own construction, with the definitions in which nobody else agrees, but which he declares to be the only correct ones.

20. *Importance of Meaning.* I want here to guard myself against the possible accusation that I am going to rely on attaching some special meaning to certain words (as metaphysicians do) in order to reach certain preconceived conclusions; or that I deceive myself by supposing that, by juggling about with words, I am altering the hard facts of the universe. This is not at all the case; on the contrary, my whole object is to avoid just such errors as these, and to bring the discussion into the closest possible contact with immediate observational experience.

The point is rather this: The way human beings behave depends largely on the beliefs, opinions, views, etc., that they hold;[1] these beliefs, etc., are formulated and communicated (if at all) in words, and influence others by virtue of the 'meanings' which these words have for the user and hearer. Moreover, the beliefs arise partly from observation, but partly also from processes known as 'thinking' or 'reasoning' on the part of the holder or (more usually) some other person; and this thinking is usually, perhaps invariably, conducted in words, whenever any but the most concrete subject-matter is concerned, and this use of words is effective only in so far as they have 'meaning'. Beliefs differ and dissensions arise, as already noted, largely because these 'meanings' are not the same for all parties; and errors may be made because the thinker supposes that a word he uses 'means' something which it does not (except perhaps to him). We can only correct such errors by finding out what the word 'really does' mean, and that involves inquiring into the nature of 'meaning' itself.

In particular, a critic who disagrees with me (or anyone else) as regards the conclusions reached in this or a similar discussion might object that he does not mean by the words 'matter' and 'mind the same things that I mean by them; or

[1] This with but little alteration is true even if we adopt a radically behaviouristic standpoint. We should then say that there is a significant correlation between the emission of noises commonly known as expressions of belief, etc., and other forms of behaviour; and would describe both as the result of certain stimulus situations, etc., and the nerve paths established by the subsequent operation of (not necessarily audible) speech mechanisms.

more flatly, that these words *do not* mean these things—but something else, in which case my contentions are invalid. If we have no theory of meaning to rely on, there is no way of telling whether his 'meaning' or mine is to be preferred, or why.

The fact that when I come to discuss situations involving what I have just called 'immediate observational experience', I shall be obliged to describe them in words, does not imply that I am thereby in any sense creating them artificially, as it were, to serve my own ends.

To sum up: No discussion can be fruitful, or the result of any abstract thinking communicable, unless all parties understand what each means by the words he uses; and it is no use asking each other what we mean by words unless we understand the meaning of 'meaning'.

Moreover, a vital part of my thesis turns on the contention that certain views now widely held, and certain theories implicitly if not explicitly entailed by them, are neither true nor false but *meaningless*, and therefore cannot compete in any way with those I advance myself. But to make good such a contention a sound and impregnable theory of Meaning is evidently prerequisite.

21. *Ambiguity of 'Meaning'*. In case it should be supposed that every schoolboy knows what 'meaning' means, the reader might profitably consider the following sentences:

1. Sericate *means* silky.
2. *Chien means* dog.
3. I did not *mean* to hurt you.
4. I did not *mean* what I said.
5. Capitalism *means* war.
6. A Universe without Justice has no *meaning*.
7. What do you *mean* by talking to me like that?
8. He is a *mean* old man.
9. He opened the door by *means* of a key.
10. The *mean* midday temperature for August 1946, in England, was about 45° Fahrenheit.

All these are current and legitimate uses of the word 'mean', etc., though perhaps it is hardly fair to include the last three, and I myself would take exception to the sixth. But it is to be noticed that the words 'mean', 'means', 'meaning', must be paraphrased differently in each case; and doubtless further examples could be found.

Suitable paraphrases would be: (1) In English, the word 'sericate' may be replaced by the word 'silky' without appreciable alteration of sense; (2) In most cases a Frenchman uses the word *chien* where an Englishman uses the word 'dog'; (3) I did not intend to hurt you; (4) When I said what I did, I ought to have said something else; (5) War is a probable consequence of capitalism; (6) No Purpose is to be attributed to a Universe in which the operation of Justice is not discernible; (7) How dare you talk to me like that?; (8) He is an ungenerous old man; (9) He used a key to open the door; (10) The average midday temperature, etc.

It does not matter whether everyone would agree with these paraphrases. The point is that we have here at least half a dozen different usages of the word 'mean', and the question is which, if any, of them is the 'right' one to rely on when we ask what a given word 'means'. Are we to select one, and say that this illustrates the meaning of 'meaning'; and, if so, what is to be the basis of our selection? Or are we to say that there is nothing that can fairly be called '*the*' meaning of 'meaning', and that, whenever we encounter the word and are doubtful about it, we must try the effect of substituting one or other of the above paraphrases, or such further alternatives as might be thought of—e.g. 'When I say 'Dog' I am referring to an animal like *this*'?

Of course, the plurality of usages indicated above may be said merely to show that the words 'mean', etc., are highly homonymous; but this does not ease the situation. It is, indeed, very important to realize clearly that our use of words is, logically speaking, entirely arbitrary and conventional, though there may be plenty of historical and psychological reasons for making one noise or mark rather than another when we wish, as we say, to refer to a particular object, etc. But the word is not a part of the 'thing', as primitive man seems to have supposed, and whatever relation there may be between them is no matter of *a priori* necessity. There is nothing whatever in logic or in law, that I know of, to prevent me asking for a pair of pink silk meanings next time I visit my haberdasher; the only objection to such a course is that, unless we have previously fixed up a private code between us, I am unlikely to obtain the pyjamas I require. To talk of the 'essential meaning' of a word, as even so sophisticated a personage as Herbert Spencer did, is to talk plain and downright nonsense. A word has no will of its own, so to speak; it does mean whatever we choose to make it mean; and Humpty Dumpty was right in saying 'It is a question of which is master—that's all'.

In these circumstances, remembering that there is nothing to be gained by too outrageous a flouting of convention (as by insisting on saying 'meaning' where other people would say 'pyjamas', or vice versa) and that 'meaning' is commonly agreed to have something to do with the relation between words and objects, etc., the only rational procedure is, I think, to examine with such perspicacity as we can bring to bear the situations and processes actually involved when words are spoken or written and heard or seen, and to define 'meaning' in terms of these processes, etc., i.e. in terms of what actually goes on. In particular, we must study the causal relationships involved; that is to say, we must inquire *why* I use one word rather than another when I wish to produce certain effects in my colloquist, and why (under appropriate conditions) the desired effects rather than others are in fact produced. In short, we must examine the whole mechanism that determines the efficacy of words, and is responsible for one word differing from another in the effect it produces.

But before we tackle this, there are a couple of matters which may conveniently be got out the way here.

22. *Dictionary Meaning*. When someone asks us what a word means, we usually reply by giving him a set of equivalent words, or by referring him to a dictionary. 'Lachrymose' means 'tearful'; 'pusillanimous' means 'cowardly', 'faint-hearted', etc. This, tantamount to defining the 'meaning' of

a word as 'the words opposite to it in the dictionary', is perhaps the most popular notion of the meaning of 'meaning'. It is all very well and extremely useful, so long as the inquirer knows the 'meaning' of the substituted words. But even within the field of a language that one would ordinarily be said to know, it may lead to perplexity. Turning up the word 'limpkin', for example, you find 'kinds of bird between cranes and rails'; and if you are no ornithologist, but acquainted with cranes only as mechanical devices for lifting heavy weights, and rails as the bars on which trains run, you are liable to be puzzled; or to take the almost legendary example, if you find 'viola' defined as 'a large violin' and 'violin' as 'a small viola', you remain uninformed until someone shows you one of these instruments and names it.

To drive the point home: imagine yourself dumped in Russia, not knowing a word of the language, no interpreter available, and equipped only with a Russian-Russian dictionary. You see a poster, say, and would like to know what it announces; but you can look up a word in the dictionary, and then those which appear opposite it, and those opposite them, etc., till the end of time, without being any the wiser. It is only when you have the luck to see, perhaps, the word 'Ruba' (or rather the corresponding Russian characters, of course) over a fish shop that light begins to dawn, and you note that 'Ruba' may mean 'fish'.

That is to say, *you can never ascertain the meaning of a word until you escape from the dictionary and make contact with the real world of observable objects and events.* The only function of the dictionary is to lead you by more or less direct or circuitous routes to some word or words, the meaning of which you know by direct acquaintance (I speak a trifle colloquially here, but not, I think, ambiguously). This is easy enough to realize in the case of a foreign language, but not nearly so easy when the language is our own, with which we have grown up and absorbed it, as the phrase goes, with our mother's milk. But it is none the less true. It is also much easier to realize where the names of objects are concerned, as in the above extreme illustration, than when we are dealing with abstractions, such as 'love' or 'loyalty' or 'confidence' where there may be quite a complex chain of sentences explaining explanatory sentences, and paraphrases of paraphrases, between the word and the observable situations which in the last stage make, as we should ordinarily say, its 'meaning' clear. But sooner or later we must invariably and inevitably come to this stage of 'pointing and shouting',[1] as Ogden and Richards call it; for if and in so far as we do not, there will always be some words left undefined except in terms of other words (which, *ex hypothesi*, we do not know), and these will remain meaningless to us.

It is very tempting to try to take a short cut and to say dogmatically that the object, situation, etc., pointed out in this last, ostensive stage—the 'this' in such sentences as 'When I say so-and-so I mean *this*—is the 'meaning' of the word; but to do so, though not very far wrong, is not quite right, and leads to serious difficulties, as we shall see in the next section.

23. *Philosophers and Meaning.* Of the Problem of Meaning philosophers have made a quite especially royal muddle, in which it would appear that they are still enmeshed. A number of the more distressing examples are given by Ogden and Richards in their book *The Meaning of Meaning*, to which I shall refer extensively below; but for a short and clear account of the lamentable predicament in which they have involved themselves I commend the reader to Ayer, *The Foundations of Empirical Knowledge*, pp. 93–7.[2] This may be briefly summarized as follows.

After exemplifying the various ways (all, I think, ostensive in the last analysis[3]) in which he suggests that the meaning of particular symbols (notably words) may be explained, he points out that 'paradoxically, this does not enable one to

[1] This is what is technically known as 'ostensive definition'.

[2] Macmillan, London, 1940.

[3] I am not quite sure whether Ayer would agree with this, though he goes so far (p. 95) as to admit that 'it is at least causally requisite that the meaning of some expressions' (symbols) 'should be explained ostensively, if any are to be understood'; but this does not matter for the present purpose.

answer the general question: What do symbols mean?' And it is this that perplexes philosophers. 'Their problem is not that they fail to understand any symbols.[1] It is of an entirely different character. It arises out of the assumption that 'meaning' is a relation which a symbol bears to something. And the puzzle is to discover the nature of this other term.' It indeed seems to have been.

Again I do not care overmuch for Ayer's wording here. Apart from the fact that 'meaning' is *not* in fact a relation between the word and something else, but the something to which the word is related, I doubt whether it would be at all sensible even to say that it is, and still more whether philosophers have assumed it; but as the last sentence quoted insists, it is the nature of the 'something' that is important.

Ayer then points out, in effect, that the natural thing to do is to say that the 'meaning' of a word or sentence is that object or situation ('empirical fact') to which we would ordinarily say that it referred, i.e. to take the line indicated at the end of my preceding section. But if we do this we come up against the difficulty that '. . . some sentences are used to express what is empirically false. Now there is no question but that these sentences are meaningful. But what they mean cannot be empirical facts; for in this case, there are no such facts. . . . If, for example, I say that there is a stove in my room and there is actually no stove in my room, I am expressing a falsehood; if there is a stove, I am expressing a truth; but the meaning of the sentence remains the same, whether or not there is actually a stove in my room. But if the meanng of a sentence is the same, whether what it expresses is true or false, and if in the case where it expresses a falsehood it cannot mean an empirical fact, then it does not mean an empirical fact even when it happens to express what is true.'

I do not like all this talk about a sentence *expressing* something which may be true or false, and would prefer to put the point more or less like this: We may agree that, to anyone conversant with the English language, the sentence 'There is

[1] It is not clear to me whether Ayer wishes to say here '. . . any particular symbols', or '. . . any symbols at all', but we may let the point pass.

a stove in my room' has an unambiguous meaning regardless of whether there actually is or is not a stove in my room. But if we try to make out that the meaning of a sentence is an empirical fact (or observable situation, etc.) then, when the sentence is what we call 'true' the observable situation is that of 'stove in room' (i.e. if we enter the room we find the stove); but if it is 'false' then the observable situation is that of 'no stove in room'. Each of these situations must be, *ex hypothesi*, the 'meaning' of the sentence; but they are different; therefore the 'meaning' of the sentence cannot be the same in the two cases. But it obviously is; therefore the observable situation, empirical fact, etc., cannot be the 'meaning' of the sentence.

What a horrible dilemma! Small wonder that philosophers have tied themselves in inextricable knots (which I will spare the reader) trying to escape from it. But the solution is relatively simple. As we shall see, the trouble is that their blushing self-effacement has led the philosophers completely to ignore the third and most important factor in the situation, namely *themselves*, as users and hearers of words, and (*a fortiori*, I suppose) such lesser mortals as may be similarly circumstanced.

It is only a mixture of primitive superstition and grammatical misdirection that leads philosophers and others to talk as if words were, so to say, active agents capable on their own account of 'meaning' this or 'referring to' that. They aren't, they can't, and they don't; so that such locutions, however convenient, are liable to be misleading. It is only the people who use or hear the words that can do the 'meaning', referring, etc.

Ayer himself admits defeat. '. . . the view I am taking is that the reason why this problem appears to defy solution is that there is really no problem to solve. We cannot find "the other term of the relation of meaning", because the assumption that meaning is a relation' (see my comments above on this) 'which somehow unites a symbol with some other unspecified object is itself erroneous.' And he concludes that 'There is no one thing that all symbols mean' (p. 98). I take it that by this last sentence he is not concerned to deny the obviously absurd proposition 'All symbols mean the same thing' (or 'All symbols have the same meaning'), but rather to deny that the meaning of any symbol may be defined in the same general terms, e.g. that it may correctly be said that 'The meaning of any symbol, S, is *x*, where *x* is some member of the class X'.[1]

The remainder of this chapter will be chiefly devoted to showing that this conclusion is incorrect, that it is possible to define the meaning of any symbol (word, etc.) in this kind of way, and to elucidating the nature of the 'something', *x*, and of the class X.

24. *The Meaning of Meaning.*[2] So far as I am aware the only worthwhile book on this subject is that by Ogden and Richards, so entitled, which I mentioned a few paragraphs back. I have had the pleasure of acquaintanceship with the authors, and have been an admirer of their work, for just over a quarter of a century, so perhaps I may be permitted to criticize to the extent of saying that I, personally, find it most irritatingly written, and in places quite gratuitously obscure; it gives me the impression, perhaps quite unjustly, that the authors are more interested in making the reader feel small and ignorant than in explaining pellucidly to him just what they want to say—but this may well be my fault. Nor, of course, do I accept every sentence as infallible gospel. But despite these animadversions I regard it as a work of the utmost importance; in fact, I do not think it too much to say that it is one of the key books of the century, and one that should not only be read but closely studied by anyone who is anxious to think clearly on any but the most concrete subject-matter.

Nearly all that I have learned about Meaning comes from this source, to which I wish to make my most grateful

[1] The supposed class X might be, for example, the class of universals, or Platonic Ideas, or other alleged 'intelligible entities', which philosophers have postulated. (H. H. P.)

[2] C. K. Ogden and I. A. Richards, *The Meaning of Meaning* (Kegan Paul, London, first edition, 1923; eighth edition, 1946. References in the text are to the fourth edition, revised, of 1936).

acknowledgements. None the less, I shall tell the story mainly in my own words.

The central feature of the Ogden-Richards Theory of Meaning (which I hold to be *the* theory of meaning, not merely *a* theory) is the point which emerged in the last section, namely, that there is no *direct* relation between a word, or other symbol, and that to which it would ordinarily be said to refer. The relation is always indirect, and mediated, so to say, by a third term, namely, the mind (or brain, as Ogden and Richards prefer) of the user or hearer. On page 11 they show a diagram in the form of a triangle, having the Symbol at the left-hand end of the base, the Referent (i.e. object, etc., referred to) at the right-hand end, and the Thought, or Reference, or 'act of reference', at the apex. There is a direct (causal) relation between the Symbol and the Thought, and another (or others), also causal, between the Thought and the Referent; but only an indirect (or, as they say, 'imputed') relation between the Symbol and the Referent. The Symbol is said to *symbolize* the Thought; the Thought is said to *refer to* the Referent; and the Symbol to *stand for* the Referent. 'Between the symbol and the referent there is no relevant relation other than the indirect one, which consists of its being used by someone to stand for a referent. Symbol and Referent, that is to say, are not connected directly . . . but only indirectly round the two sides of the triangle.'[1] The only exceptions to this rule arise in the case of onomatopoeic words, mimetic gestures, representational symbols such as drawings, etc.; but these need not concern us here.

Strictly speaking, the 'meaning' of a symbol *is* the 'thought' or 'reference' or 'goings on' (as O. and R. have it) in the brain and nervous system of the person involved in the semantic situation, and it is necessary to use the word in this strict sense in order to answer the philosophers' conundrum discussed in the previous section; but for most practical purposes, as we shall see, it is harmless enough (provided always that we know what we are doing) to short-circuit the triangle, so to

[1] *The Meaning of Meaning*, pp. 11–12.

say, and to take the refer*ent* as being the 'meaning'—as is more or less common practice.

All this will become quite clear, I think, when we have examined the fundamental processes as a result of which words come to be connected with thoughts, etc., and through them with objects and events. To do this we must begin with the general theory of signs; for words and other symbols are only signs of a particular kind, namely, those used more or less deliberately by human beings for purposes of communication and thinking.[1]

25. *The Meaning of Meaning, contd.* (2)*; Signs.* Situations involving signs, whether we recognize them as such or not, are very much commoner than is usually supposed. 'Throughout almost all our life we are treating things as signs. All experience, using the word in the widest possible sense, is either enjoyed or interpreted . . . or both, and very little of it escapes some degree of interpretation' (O. and R., p. 50).[2] And interpretation is synonymous with treatment as a sign.

It is difficult formally to define a sign without considerable complication of wording, so I think it will be best to get at the definition by degrees.

Intrinsically, a sign may be of any character whatever; it may be '. . . any stimulus from without, or any process taking place within' the organism (O. and R., p. 52); for example, we take a red sky at sunset to be a sign of good weather to come, or the reddening of litmus paper as a sign of acidity, and so forth in innumerable cases. Everyone will agree that, in such situations, we are 'interpreting' 'signs'; and also that we do so by virtue of past experience in one form or another. What has happened is that we (or others before us who have passed on their experiences) have observed that a red evening sky is usually followed by a fine day; and generations of chemists have found that solutions, identified as 'acid' on other grounds, have turned litmus red. But for these prior

[1] We need not trouble ourselves here with such processes of communication as occur among the lower animals, or with specialized usages of the word 'symbol', as, for example, by the psycho-analysts.

[2] I shall give references to Ogden and Richards's work in this abbreviated form when convenient.

experiences, direct or by proxy, we should be unable to interpret the signs; they would, as we say, have no 'meaning' for us, and the evening sky might as well be green, or the litmus turn yellow, for all the information we should gain from the phenomena.

Note that prior experience of the sign itself is not enough; it is essential that it (or something very like it) should have been encountered in a context or contexts more or less similar to that in which it is re-encountered. In the first instance the stimuli constituting the sign form part of a larger group, or 'stimulus situation'. To this situation the organism reacts, in some way or other—not necessarily appropriately. The stimulation and reaction leave traces, which we may describe in terms of Semon's 'engrams',[1] or otherwise, according to taste; and the organism is then said to be 'adapted' to that situation. If the situation recurs, even in part, there is a tendency for the whole of the residual traces to be re-excited, and for the organism to react again in the same way as on the first occasion. Thus, if the sign-constituting stimulus recurs in a context of other stimuli sufficiently like that in which it was first encountered,[2] it will tend to evoke the adaptations made by the organisms to that first context as a whole.

To abbreviate slightly an example from Ogden and Richards: Professor Lloyd Morgan's chicken seizes a yellow and black caterpillar, finds it (presumably) offensive in taste, and rejects it. 'Thenceforth the chicken refrained from seizing similar caterpillars. Why? Because the sight of such a caterpillar, a part, that is, of the whole sight-seize-taste context[3] of the original experience, now excites the chicken in a way sufficiently like that in which the whole context did, for the seizing at least not to occur, whether the taste (in images)

[1] See R. Semon, *The Mneme*, tr. by L. Simon (Allen & Unwin, London, 1921), Ch. II.

[2] How great a part of the situation must recur in order to ensure this, or how close must be the similarity between the various occurrences and recurrences, is a question of fact to be settled by observation in particular cases.

[3] Note that this must include the kinaesthetic sensations (stimuli of endosomatic origin) accompanying the motions of seizing, etc. W. W. C.

does or not.' (O. and R., pp. 52–3.) The stimulus yellow-and-black, at least in the relevant setting, has become a *sign* of offensiveness in taste, which the chicken may be said to *interpret*. Similarly, as already noted, certain shapes and colours of clouds in the sky may be *signs* of rain to the weather prophet; a high temperature and spots on the chest *signs* of measles to the physician; persistent yawning a *sign* of boredom to the raconteur; and so on in innumerable situations. But the mechanism whereby the various stimuli acquire efficacy as *signs*, as opposed to being just stimuli of no special interest, is invariably the same.

The observer initially encounters a stimulus (or relatively small group of stimuli), S, in a certain context, C, (i.e. as forming part of a larger group), and this context includes some other stimulus, R, say, or is shortly followed by one, C′ say, which does. As an organism he reacts in some way or other to the situation, and the process produces certain changes in his nervous system, brain, or mind; he is then said to be adapted to the situation or context C (including C′ in relevant cases). On some subsequent occasion he again encounters S, or some substantially similar stimulus, S′, in the same or some substantially similar context, C″; this re-excites the traces, etc., originally left by C (and C′), as a whole, including that left by R, so that he tends to react to the total situation in the same way that he did before, notably including his reaction to R. In the ultra-simple case of the chicken, his instinctive reaction to the stimulus-situation 'crawling caterpillar' was to peck and seize; but his reaction to the immediately following stimulus R (nasty taste) was to reject (open beak and drop). On the next substantially similar occasion he tends to react in both these ways; but he cannot both seize and reject, so that the two incompatible reactions cancel each other out (or, in general, tend to do so), and the seizing reaction is, as we say, 'inhibited'. Similarly, the weather prophet says 'Rain soon' and runs for his umbrella; the doctor looks wise and diagnoses 'Measles'; the chemist notes 'Acid' and proceeds accordingly.

Since the interpretation of the sign depends, in all cases,

on their having been previously encountered in similar situations, in conjunction with other stimuli, it goes without saying that, if the R-elements, so to call them, of these situations had been different, the interpretations would have been different also. Needless to say, too, in the case of human beings who have speech and writing at their disposal, the previous encounters may have been vicarious: the doctor may merely have read that high temperature and spots on the chest are signs of measles; the chemist may only have been told that reddening of litmus is a sign of acidity, without having himself carried out all the experiments which enabled his predecessors to classify certain solutions, etc., as 'acidic', and to correlate the reddening of litmus with these. But the original sequence of events must have been the same for someone; and it is, as we shall see in a moment, only by virtue of words acting as signs that such information can be communicated.

The foregoing is, I think, a substantially correct account of the Theory of Signs, as expounded by Ogden and Richards,[1] and may conveniently be summarized by the following quotations from their work.

'A sign is always a stimulus similar to some part of an original stimulus' (stimulus situation) 'and sufficient to call up the engram formed by that stimulus' (or 'to call up an excitation similar to that caused by the original stimulus'). 'An engram is the residual trace of an adaptation made by the organism to a stimulus. The mental process due to the calling up of an engram is a similar adaptation: so far as it is cognitive, what it is adapted to is its referent, and is what the sign which excites it stands for or signifies.' (O. and R., p. 53.)

'Our interpretation of any sign is our psychological reaction to it, as determined by our past experience in similar situations, and by our present experience.' (p. 244.) '. . . the peculiarity of interpretation being that when a context' (or stimulus-situation) 'has affected us in the past the recurrence

[1] In saying this, I do not wish to suggest that there might be an alternative and equally plausible theory held and expounded by someone else. This is, I submit, *the* only tenable theory, i.e. the only one compatible with the facts of common experience and psychological study.

of merely a part of the context will cause us to react in the way in which we reacted before.' (p. 53.) '. . . to be an act of interpretation is merely to be a peculiar member of a psychological context of a certain kind; a psychological context being a recurrent set of mental events peculiarly related to one another so as to recur, as regards their main features, with partial uniformity.' (p. 57.)

Note in passing that although the foregoing account savours considerably (to my mind at least) of Behaviourism—an attitude to which I, personally, consider that the authors incline unduly—there is nothing to render this obligatory. An account which would, I think, show an exact point-to-point parallelism could equally well be given in terms of 'mind', 'sensa', 'images' (with due caution), 'association', etc.

26. *The Meaning of Meaning, contd.* (3)*; Words as Signs.* Now let us turn to *words*, which, together with certain other symbols (e.g. mathematical), are only a particular sort of sign. As Ogden and Richards say, 'When we consider the various kinds of Sign-situation . . . we find that those signs which men use to communicate with one another, and as instruments of thought, occupy a peculiar place. It is convenient to group these under a distinctive name; and for words, arrangements of words, images, gestures, and such representations as drawings or mimetic sounds, we use the term *symbols.*' (p. 23.)

If we consider the history of our personal relation to any particular word we shall find that, in simple cases, it is something of this kind. In our infancy we encounter various *objects*, as we may call them, in various contexts. More strictly, we encounter various groups and sequences of stimuli which arouse sensations, each group having a certain degree of coherence. Many of the contexts in which these groups are encountered include other stimuli, usually in the first instance auditory. Thus the groups constituting experience of the *object* Chair, say, i.e. of seeing, touching, bumping into, sitting on, falling off, etc., form parts of larger groups (contexts or situations) of which in many cases the peculiar noise of the spoken *word* 'Chair' also forms a part. This is exactly the condition under which one stimulus or stimulus-group, to wit the noise 'Chair', may become a *sign* of another stimulus-group, to wit the *object* Chair; and the word (noise) 'Chair' does in such circumstances become in fact a sign of the object Chair, in precisely the same way and for precisely the same reason that the cloud-shapes, spots, and reddened litmus become signs of rain, measles, and acidity; so that, when we hear the word 'Chair', the 'residual traces of', or more generally the effects left, by previous situations including this stimulus-group are re-excited, and we take up, in greater or less degree and more or less overtly, our adaptive attitude to situations of a chair-containing type.

We may then be said to know the meaning of the word 'Chair', and this state of adaptation, physiological or psychical or both, *is* the meaning of the word 'Chair' for us.

The last two words are not without importance. If our parents had been so unkind as habitually to have emitted the noise 'Pig' whenever ordinary people would have emitted the noise 'Chair', the word 'Pig' would have come to mean for us what the word 'Chair' means for normally conditioned persons; and there would presumably have arisen some confusion in later life, and some difficulty in readjustment. Incidentally, malconditionings of this kind, less extreme but more subtle and insidious, are probably responsible for many more conflicts and consequent neuroses (I use the word untechnically) than is commonly realized. Mention of such a possibility, by the way, should serve to emphasize the point that there is, and can be, no unique, *a priori*, God-given meaning that a word *must* have: it means whatever circumstances have in fact made it mean to whoever may be concerned. The process whereby we learn the meanings of words other than the names of familiar objects, colours, tastes, etc., or of simple activities may, of course, be highly circuitous and immensely complicated; but however indirect it may be, and however many stages there may be of encountering new words or phrases in conjunction with verbal forms, of which the meaning is known by their having been experienced in conjunction with other verbal forms, and so forth, the basic mechanism remains unchanged, though naturally the possibilities of going astray and of divergences of usage developing as between different people are enormously increased.

We can now see clearly how it comes about that different words have, in general, different 'meanings' (in the sense indicated above) for any given person—and sometimes, even, the same word for different people, according to their experience. The determinative contexts in which the word 'Chair' is encountered will (in normal circumstances of upbringing, etc.) be different from those in which the words 'Fire', 'Cat', 'Tomato' correspondingly figure; so that the 'meanings' of these words will come to be different from that of the word 'Chair' and different also *inter se*. Since this basic process is, colloquially speaking, the only way in which any word can come to produce a characteristic effect, or to be differentiated otherwise than as regards its intrinsic sound or shape from any other word, it seems to me that it would be merely perverse to use the word 'meaning' in any sense other than that so given; though, of course, it is *logically* permissible to reserve some other word, such as 'significance' to refer to the outcome of the processes just described, and to use 'meaning' as synonymous with 'purpose', 'intention', etc., as in fact it is also used—and harmlessly so, provided we remain aware that these usages are homonymous in character. But the above is what makes words work; and that is what matters.

So far, so good; and if anyone feels inclined to complain that I have spent a great deal of time elaborating the obvious, I can only reply that, although it seems obvious enough as soon as stated, no one except Ogden and Richards, so far as I know, has thought it worth while mentioning in connexion with the Problem of Meaning before; and that, if philosophers had paid a little more attention to the obvious, and a great deal less than they have[1] to puzzling their heads over 'unique relations', 'propositions', and the rest of it, we should

[1] Cf. Ogden and Richards, loc. cit., Ch. VIII. [But cf. also Berkeley, *Principles of Human Knowledge*, Introduction, Sect. 18 and following. Berkeley, for one, did 'pay attention to the obvious' in this connexion. (H. H. P.)]

never have got into our present tangle. Incidentally, there would have been no metaphysics.

But there is a good deal more to be said yet.

27. *The Meaning of Meaning, contd.* (4); *Reference.* In the very earliest stages, it seems reasonable to suppose that the word is a sign to the child of the 'thing' or actual object of which it is the name,[1] especially when he is being deliberately taught the names of objects; and that when he hears the word he tends to react as if in the presence of the object, or at least enjoys visual, tactile, etc., images of one kind or another such as we would ordinarily describe as images 'of' the object. I do not think it matters for our present purpose whether this surmise is correct or not, for there can be no doubt that this stage is very soon overpassed in normal people.

Children appear to be naturally imitative, and their tendency to pronounce the name of an object which they have, as we say, 'learned', whenever they see that object (and not infrequently *ad nauseam*), is a matter of common observation. But they soon find that it is quite possible to pronounce the name when the object is not corporeally present; moreover, they often hear the word (name) under like conditions. Of what, in such circumstances, are we to say that the word is a sign, and what may it be said to 'symbolize'? I do not think anyone can quarrel with the statement that it is a *sign* that the child, or other person, is—to put it colloquially—'thinking about' the object of which the word is the name. It is of no consequence whether we describe the state concerned in terms of the re-excitation of traces, the stirring up of engrams, or the revival of images by association; what matters is that the enunciation (or, at a later stage, the writing, etc.) of the word is a sign of the resuscitation of the condition—neuro-cerebral, mental, psychical, or what you will—which constituted the adaptation of the individual to the situations in which he has encountered the object and word in conjunction, as modified, of course, by any subsequent experience that may have been relevant, and by the contemporary situation.

[1] Compare what we said above about the name being identified with, or regarded as part of, the thing by primitive peoples. (Ch. II, pp. 33–5.)

This condition is what Ogden and Richards call an 'act of reference', or, more briefly, 'reference', though I shall occasionally refer to it when convenient as a 'state of mind', without prejudice to the terms into which this phrase should be expanded.

It cannot be too clearly realized that this condition, 'act of reference', 'reference', or 'state of mind', *is* the meaning of the word (or other symbol, of course) *for* the enunciator, *on* the occasion in question. It will not in general be *quite* the same for any other enunciator, or for the same one on any other occasion.

When the child or other person *hears* the word, it induces in him an act of reference or state of mind which is similarly dependent on the situations in which *he* has encountered the object and word in conjunction, modified as before, and this reference *is* the meaning of the word for *him* on the occasion concerned.

Since the experiences of no two people will ever be quite the same, these two references will never be strictly identical; and the same word will never, strictly speaking, mean quite the same for any two people. Sometimes this gives rise to a great deal of trouble, as when Mr. Bevin and Mr. Molotov each talk about 'democracy' with, evidently, very different references arising from very different experiences of political organization, etc. But even a discussion of cats might end in misunderstanding, if your experience were derived mainly from tiger-cats or polecats or even from Siamese, and mine from the less esoteric domestic varieties. And to take an extreme case, so unambiguous a symbol as π (even when so written) will presumably excite references flavoured, so to say, in varying degree in different people by association with Pie (apple), Pie (pork), or Pie (in the sky).

On the face of it, communication would appear to be a hazardous and error-fraught process; and so it is, so soon as we start talking about anything that cannot be unambiguously identified, if need be, by ostensive demonstration, or using words which cannot be replaced satisfactorily by others amenable to similar treatment. It is here that considerations of Common Usage, Good Use, etc., come to our aid, together with the fortunate fact that our experiences of most subjects of discussion, though never identical, are sufficiently similar to ensure a considerable overlap in the references we make when using or hearing, etc., the relevant symbols.

In the case of the mathematical symbol π, we may fairly say that the overlap is virtually complete, at least in mathematical contexts. If you and I are discussing cats, it is sufficient for most practical purposes; although your experience of cats will not be quite the same as mine, it is likely to be sufficiently similar to prevent misunderstanding on most points, since all ordinary cats have characteristics in common which are numerous compared with those as to which they differ. Even the references excited in Mr. Bevin and Mr. Molotov by the word 'democracy' presumably both include the idea that, in a democratic state, all the people (theoretically and in principle, at least) have some say in the affairs of government, however much they may differ outside this common area.

In other words, within the area covered, so to say, by the assorted experiences of various people in respect of any given subject-matter, there is what I may term a certain 'hard core' of common experience, and a corresponding core, therefore, common to the references made by them when hearing or uttering a given word—provided, of course, that they are familiar with the usage of the language concerned. It is this, and this alone, which makes communication possible, though at the same time it limits its perfection.

28. *The Meaning of Meaning, contd.* (5); *Correctness, etc., of Symbols.* Neglecting for the time being the cynical view that 'the object of speech is to enable us to conceal our thoughts', we may reasonably say that the function of any communicatory process is to enable others to share, so far as may be, our thought in respect of whatever subject-matter is under discussion—or words very much to this effect. The object, that is to say, of using a particular word, or one of those more complex symbols known as sentences, statements, or propositions, is to induce in our hearer a state of mind, or

reference, similar in the relevant respects to our own at the moment of speaking. In so far as the symbol we choose to employ does in fact achieve this result, it is successful, and may be termed *correct*. 'A symbol is correct when it causes a reference similar to that which it symbolizes, in any suitable interpreter.' (O. and R., loc. cit., p. 206.) 'An incorrect symbol is one which, in a given universe of discourse, causes in a suitable interpreter a reference different from that symbolized in the speaker.' (p. 102).[1]

But a *correct* symbol is not necessarily a *true* one. Failure to realize this is responsible for the philosophic woe mentioned in section 23 above, as we shall discover shortly. The symbol may be perfectly correct, in the above sense of inducing in you a reference or state of mind substantially similar to my own; but if my references are false, so will yours be. If you ask me the way to the nearest pub, and I reply, 'Keep straight on till you get to the cross-roads, and then turn to the right', this instruction may be misleading for either of two reasons. First, I may correctly visualize, etc., the relevant geography, and be well aware of the distinction between 'right' and 'left', but may none the less inadvertently and by a pure *lapsus linguae* say 'right' by mistake, whereas the local dispensary is actually to the *left*. In this case my reference is true, but my symbol is incorrect, and *false*.[2] Secondly, I may visualize, etc., the geographical situation wrongly, e.g. I may only have approached the hostelry in the past from some other direction, and be obliged to work out in my mind how to get to it from where we stand—in this case my reference is *false*, and so is my symbol, though it is perfectly *correct*, in that it duly induces in you a reference (visualization, etc.) similar in the relevant respects to my own. In either of these cases you are liable to go thirsty.

[1] 'A universe of discourse is a collection of occasions on which we communicate by means of symbols.' (Ibid., note.) Any such collection is characterized by at least an ostensible likeness of subject-matter and point of view, etc.; as when we discuss Poles (hop), *or* Poles (from Poland) *or* Poles (geographical). And a suitable interpreter is one who is familiar with the common usage of the language in the given universe of discourse.

[2] It should go without saying that the whole sentence is merely a complex symbol as noted above.

In general a symbol (notably a statement or proposition, etc.) is true only if it is itself correct, and the reference it symbolizes is true, or, as Ogden and Richards prefer to say, 'adequate'. If either the symbol is incorrect, or the reference it symbolizes is false, then the symbol will, in general, be untrue. In the special case considered, you will as a matter of fact get your drink (assuming there to be any available) if my reference is false (i.e. I imagine your objective to be to the right instead of to the left) *and* my symbol is incorrect (i.e. I erroneously say 'left' when the correct symbolization of my reference would be 'right'). But this is only because we have presupposed that there are only two mutually exclusive alternatives.

Note in this connexion that if, disapproving of your presumed intemperate propensities, I deliberately mislead you by saying 'right' when I know perfectly well that the tavern is to the left, then it is a case of incorrectness of symbol. My reference must be true (at least to the extent of suspecting that 'right' is misleading), otherwise I could not know how to misdirect you; and the *reasons* for an incorrect symbol being used have nothing to do with the *mechanism* of communication.

29. *The Meaning of Meaning, contd.* (6); *True and False References.* We must now inquire into what it is that determines whether a reference is 'true' or 'false'.[1] To put it very roughly, as a sort of first approximation, we may say that a reference is true if that to which it refers (the refer*ent*) is in fact of the kind that the reference takes it to be, but false if it is of a different kind. As we have seen, a reference is always an adaptation of the organism, brought about by a stimulus-situation of which one or more components have occurred before and act as signs; alternatively, it may be described as an interpretative state of mind.[2] The referent is that which

[1] Cf. Ogden and Richards, loc. cit., p. 62: 'If . . . there be an event' (discoloration of blotting-paper) 'which completes the external context in question,'—viz. a genuine spilt-ink situation, expectation of which constitutes the interpretation of the sign 'shiny, dark-blue patch'—'the reference is *true* and the event is its referent. If there be no such event the reference is *false*, and the expectation is disappointed.'

[2] We shall see in the next chapter that all perceptual situations are essentially of an interpretative character.

the organism is adapted *to*, or that which the signs are interpreted as being *of*. *Mal*adaptation and *mis*interpretation are by no means excluded from possibility by any natural law, and are in fact common enough. For example, someone plays a practical joke on me by arranging one of those celluloid ink-pools and an overturned ink-bottle on my writing table. Being of a naturally naïve and unsuspicious disposition, I interpret the visual signs (irregular, dark-blue, shiny patch, etc.) as being of a genuine spilt-ink event; my adaptation to the stimulus-situation, including these previously experienced signs is to grab the blotting-paper; and I symbolize my reference (interpretation or adaptation, so far as it is cognitive) by exclaiming, 'Drat that cat—she's upset the ink!' In such circumstances I have misinterpreted the signs, my adaptation is inappropriate, and my reference is false. How do I (subsequently) know this? Because the spilt-ink situation to which, on the basis of past experience, I am adapted, is not, so to say, fulfilled. When I cautiously advance the blotting-paper to touch the surface of the supposed ink-pool, it encounters a premature resistance and slithers over it; and there is no spreading discoloration such as I expect.

Are we then to say that the spilt-ink situation 'does not exist', with the corollary that my reference is to nothing, or to a 'non-existent entity', or (in the worst manner of philosophese) to an entity that 'subsists' but does not 'exist'? I think not. Certainly the spilt-ink event does not exist in the physical world, for part of the definition of ink is, in effect, that it is absorbed by blotting-paper (i.e. if it is not, then it is not *ink*, within the meaning of the act). But it exists all right in the form of the traces, etc., left in me by previous experiences of similar, physically existing, events, or of the images which the sight of the artificial pool may evoke; in short, we may reasonably say that the spilt-ink event exists in the realm of my imagination, though not in the world of fact. And imaginary events, etc., are just as real in their own way as physical events; for, as I shall have occasion to stress in due course, all this talk about 'reality' is just so much nonsense.

In other words, what is wrong with my reference in such a case as this is that it is, so to say, misplaced. My true reference is to past physical and present imaginary events; it is false only in so far as it is extrapolated, so to say, to present or immediately future events of the physical world. If instead of jumping to conclusions and falsely accusing an innocent animal I had interpreted the signs more cautiously, and had been content to say 'It looks as if ink had been spilt', or 'I think ink has been spilt', all would have been well; my reference would have been to a situation imagined in my own mind, the referent (this imagined situation) would have been properly located, so to say, and my symbol would have been not only correct but true.

Similar considerations apply to the objects, events, situations, etc., of hallucination, dream, or imagination of a more deliberate kind. It is no use contending that hallucinatory, oneiric, or imaginary objects do not exist, unless you start by imbecilically synonymizing 'exist' with 'be material'. Such objects (pink rats, phoenixes, pea-green centaurs, etc.), built up, so to say, of what we may momentarily call 'images', unquestionably exist, and their existence can be verified—provided you look in the right place, viz. not in the material world but in their own 'mental' sphere. The symbol 'King Charles I died in his bed making witty remarks' (cf. O. and R., p. 102) may obviously be *correct*, inasmuch as the speaker may be historically misinformed; but it cannot be *true* unless the referent of the reference symbolized is 'placed' (e.g. by the literary context) in the world (say) of the day-dreams of a perfervid Jacobite, and not in that of historical events.

Unless a symbol is so expanded as unambiguously to 'place' the referent of the reference which it symbolizes, it is futile to discuss whether it is 'true' or not. Usually this precaution is omitted, and it is taken for granted that the place claimed, so to say, for the referent is in the order of physical events; but this is by no means necessarily the case, and much confusion accordingly arises from the omission. If I say, 'There was a lobster in my bed last night, and he pinched my toe', the statement will ordinarily be taken, if unqualified, as symbolizing a reference to an event in the physical world, and would normally be received with surprise approaching incredulity, though it could, of course, be true. But if I dreamed the incident, and say 'I dreamed that there was a lobster . . . etc.', then the symbol correctly symbolizes a true and 'adequate' reference to a situation in the world (or 'realm', 'order', etc.) of dream life, in which a dream-lobster dream-pinches a dream-toe, and all is well; the referent is both characterized and 'placed'.

This adequate placing of referents in their 'order'—of historical, physical, dream-imaginary, wish-thinking, etc., events—is of the utmost importance, for otherwise we cannot tell what steps to take to verify them and so to ascertain whether the symbol is true. But it would be a mistake, as Ogden and Richards duly emphasize, to suppose that places and orders are like the blank squares of a crossword puzzle, which may or may not be filled in. The term 'place' is 'rather a symbolic accessory . . . than an actual symbol' (O. and R., p. 106); that is to say, it does not stand for or refer to anything over and above the referent itself. If we know all about the referent, we know all about its 'order'. 'When we say that a referent is allocated to an "order"' (i.e. is 'placed', W. W. C.) its "order" is short-hand for those parts of the reference by the aid of which we attempt verification.' (O. and R., p. 292, note.) 'There is no difference between a referent and its place. There can be no referent out of a place, and no place lacking a referent. When a referent is known its place is also known, and a place can only be identified by the referent which fills it. "Place", that is, is merely a symbol introduced as a convenience for describing those imperfections in reference which constitute falsity.' (pp. 106–7.)

30. *Meaning and the Problem of Truth.* We are now in a position to deal with the philosophers' dilemma of section 23.[1] This arose, as we saw, from supposing that the 'meaning' of a sentence is an empirical fact. This involved the difficulty of explaining how it comes about that a sentence, which can

[1] pp. 51–4. above.

obviously only have one meaning, whether it be true or false, is none the less connected with either of two different observable situations, according to which it is.

The solution to this Sphinxian riddle is obtained by the more rational and more detailed analysis of sign situations sketched above. According to this, the meaning of a symbol (e.g. of a sentence, which is a complex symbol) is *not* an empirical fact, but a psychological reaction, adaptation, state of mind, or reference; and we have seen that this may be true or false ('adequate' or 'inadequate') according as it does or does not place the referent in its proper order, regardless of whether it is correctly symbolized by the symbol used.

Thus, if I say 'There is a stove in my room', this symbol may be supposed to be 'correct', in the above-indicated technical sense that it produces in a suitable interpreter (i.e. anyone familiar with the English language, in this case) a reference similar in all essentials to that which I am making; and these references are its meaning,[1] and this is single, as is required. But there are at least two 'places' which the referent of this single reference may occupy—one in the order of physical events, the other in the order of (say) my beliefs (imaginings, etc.). Alternatively we may say that the same place, in the order of physical events, is claimed by two different referents, to wit 'Stove in room' and 'No stove in room' (or equivalent phrases).

To ascertain whether the symbol is true or false we go and search the room. If we find a stove, we say that the order of physical events does include the referent 'Stove in room', and the symbol is true; if not, we say that the place in the order of physical events claimed by the referent 'Stove in room' is occupied by the referent 'No stove in room' and that the referent 'Stove in room must be sought in some other order; and the symbol in this case is false. For a more detailed treatment, see Ogden and Richards, op. cit., Appendix E.

Note that the way philosophers try to get out of their

[1] Strictly, as already pointed out, there are two not quite identical references; but we take their 'overlap'—which in such a case will be large—or 'highest common factor', so to say, as a single meaning common to both.

trouble is by inventing a class of would-be facts, which they call 'Propositions' (I have given the word a capital letter to avoid confusion with 'propositions' in the more correct sense of complex symbols; besides, it appropriately suggests their wholly mythological character. 'By doing this,' says Ayer[1] 'they are able to provide a verbal solution to the problem' (i.e. by saying that though the sentence has a single meaning, the Proposition may be true or false) '. . . but the solution is no more than verbal. We are told that what a sentence means is a proposition' (of the capital P sort); 'but if we then ask what a proposition is, the only definition available is that is what a sentence means.'

Small wonder that 'philosophers who regard themselves as empiricists should find themselves unable to attach any significance to this notion of real propositions' or even that at the last '. . . they fall into the error of formalism'[2] and (one must suppose) forthwith perish miserably.

But it isn't a mythological and meaningless 'Proposition' which is either true or false; it is the *reference*, which is a perfectly respectable set of existents—resuscitated images, excited engrams, or what you will.

Once the Theory of Meaning is properly grasped, there is no need at all to perplex ourselves by talking about a symbol having a single 'meaning' which none the less somehow contrives to be dual.

At the same time, and as a matter of general interest, we can answer Pilate's famous question 'What is Truth?' The trouble has arisen through the all too common habit of hypostatization, that is to say, of assuming that because there is a word in the language there must somewhere be a 'thing' or a 'substance' in the universe to correspond to it. Thus Parmenides declares 'Thou canst not know what is not—that is impossible';[3] and as we are quite sure that we occasionally know 'the truth', there must be something called 'the truth'—and more generally 'Truth'—to know.

In fact, of course, that there is no such thing as Truth;

[1] *The Foundations of Empirical Knowledge*, pp. 96–7. [2] Ibid., p. 197.

[3] Requoted from Russell, *Our Knowledge of the External World*, first edition, p. 166.

there are only true propositions; and the proper answer to Pilate's question is, 'Truth is a symbol, which enables us to refer to other symbols, known as propositions, by means of which facts are characterized and allocated to their order'.

This is rather a test case, for if the reader understands why it is correct, he may fairly be said to understand the Theory of Meaning; but if not, then not. Most people tend to boggle at the flat statement 'Truth is a symbol' and want either to say 'The *word* "Truth" is a symbol . . .', or to adopt some other subterfuge which will enable them to keep some mythological entity alive, so to say, in the background. But Truth *is* a symbol; that is to say, it is a mark on the paper like any other (here it is—Truth—look at it); and to say 'The truth is that (so-and-so)' is no more than an exact equivalent of saying 'The proposition (so-and-so) is true'. And this in turn is equivalent to saying that some fact or event is duly allocated to its order by the reference which the proposition correctly symbolizes. Or if we say 'Truth is hard to find', this is only an alternative way of saying that it is often difficult to decide the order to which the referent of the reference symbolized by a proposition should be allocated. And similarly for other locutions. So let us have no more nonsense about Truth, which is only a convenient shorthand symbol, but confine ourselves in this connexion to considering the evidence bearing on whether particular propositions are true; that is to say, on whether the referents of the references which they symbolize are to be found in the 'places' to which they have been allocated.

31. *Private and Public Meaning: Meaningless Symbols.* We should now have a pretty clear idea of what we mean by 'meaning', but we should none the less think twice before too lightly speaking of '*the*' meaning of a symbol. Strictly, as I have already remarked, we should only speak of *the* meaning of a symbol in connexion with a specified person on a specified occasion. 'The meaning of any symbol S for person P on occasion O (or at time T, or in context C, which come to the same thing) is the reference R (or psychological reaction, or state of mind, or adaptation, etc., in so far as these are

cognitive) caused by or causing, as the case may be, the hearing or uttering of that symbol by that person on that occasion.'[1] But, as I have remarked, there is as a rule sufficient overlap or common core in the references made by different persons, or by the same person on different occasions, to make discussion practicable, so that the inter- and intra-personal variations may usually be ignored.

The other important concession commonly made to convenience is to omit the reference altogether, and to say that the meaning of a symbol *is* that which, in the Ogden-Richards terminology, it is said to 'stand for', i.e. the referent of the reference which it symbolizes. That is to say, we short-circuit, as it were, the two direct-relation sides of the triangle,[2] and use the 'putative' relation between the symbol and the referent. That this may be exceedingly dangerous is shown by the whole history of linguistic disputes, and particularly by the tangle into which philosophers enmeshed themselves, as explained in sections 23 and the preceding. But for most practical purposes, as opposed to theoretical inquiry, it is harmless enough, and an indispensable aid to readiness of communication, which would be hardly practicable without it.

The rather nice question now arises of whether and, if so, in what sense, it is legitimate to describe the characteristic words, phrases, and propositions of metaphysics as 'meaningless', instead of merely saying that they are false (or conceivably true).

Speaking quite strictly, it clearly is not; for, on the theory of 'meaning' presented above, it is necessary to suppose that any symbol or other sign produces *some* sort of psychological reaction in anyone who encounters it; and this, by the definition adopted, is its 'meaning' for that person at that time. Any scrawl or noise, however 'senseless', is bound to be vaguely reminiscent or suggestive of something, and may even well be suggestive of much the same sort of thing to

[1] This is, of course, the 'one thing which all symbols mean' as opposed to Ayer's view that there is nothing of which this can properly be said. Cf. Sect. 23, above, pp. 53–4.

[2] See p. 55, above.

different people. For example, the mainly nonsensical sentence (O. and R., p. 46) 'The *gostak distims* the *doshes*' is not wholly meaningless in this sense The word *gostak* vaguely suggests 'goshawk' and Gosport, *distims* ought, one feels, to have something to do with distilling and *doshes* with goloshes. Such words do induce references of sorts which may fairly be said, so long as we keep to the strict letter of the law, to constitute their meanings.

Similarly, the words used by metaphysicians doubtless induce or are caused by more or less characteristic states of mind in those who hear or utter them. Even a metaphysician will not, in general, use the word 'reality' when his reference is to cheese; and few will react with violence to the statement 'I regard you as essentially a Continuant'. We may even concede that, at the dictionary level, so to say, metaphysicians may be quite precise and consistent in their usage of technical terms—in the sense, that is to say, that each will always react with the same alternative form of words if asked what he means by one of these terms; and cases are not unknown in which two or more metaphysicians may react indistinguishably to such questions, though this is rare.

But if by persistence or brutality we succeed in chivvying the metaphysician from between the covers of the dictionary, and demand that he should so characterize and place his referent that we can take steps to ascertain whether his placing is correct, we find that he cannot do so. The words he uses—Absolute, Reality, Essence, Soul, Deity, Continuant, etc., etc.—have no referents other than the verbal and other images, or engram excitations, etc., within his own mind or body; and the 'place' of the referent is, so to say, purely private.[1]

The propositions and conclusions of metaphysics, however,

[1] Perhaps I ought, in strictness, to make here a very tentative reservation to the effect that, as the subsequent course of the discussion will show, it seems not theoretically impossible that these images, if common to many minds, might come to form 'synthetic' (mental) objects of a sort, having a certain measure of stability. If so, then these would be the relatively 'public' referents of the words; but they would not at all be the referents to which metaphysicians suppose themselves to be referring by the words they use.

are supposed to be concerned with the public world in which we all live and move and have our being, not confined to one formed solely of the metaphysicians' own imaginings, mental states, etc.; and it is waste of time for a metaphysician to declare that 'for him' or in his private world of thought some such proposition as 'The Absolute is incapable of evolution' is true. It may very well be that the form of words which for him is equivalent to the term 'The Absolute' is such as to be incompatible with those which for him are equivalent to 'capable of evolution'; but this has nothing to do with the public universe except that it happens to be a trivial phenomenon in it.

The crux of the matter is reached when he is asked how he proposes to ascertain whether such a conclusion as this is true; and we find that to such a challenge he can give nothing but a purely verbal answer—he cannot say 'Do this, and you will observe that' or give any similar instruction that will throw any light on the question at all. It is impossible, *ex hypothesi*, to make any observations at all on the Absolute; there is accordingly no identifiable referent for which the symbol can be said to stand; this is indistinguishable from there being *no* referent in any public sense; and 'Whenever a form of words has no referent, it ceases to be a symbol and is nonsense'. (O. and R., p. 292.)

That is to say, no symbol is utterly meaningless in the sense of having no causally related reference of any sort,[1] but it may be so (and those typically used by metaphysicians are) if we adopt the usual 'short-circuit' definition of 'meaning' as the (assumedly public) referent of the reference, which the symbol is commonly said to 'stand for'.

It is accordingly legitimate to say that a word is *meaningless* when it is impossible to give instructions for finding the referent for which it is alleged to stand; or that a proposition is meaningless when it is similarly impossible to give instructions for obtaining data relevant to its truth or falsity, which

[1] I do not think we need bother ourselves here with the question of whether symbols which have never yet been used are exceptions to this assertion.

is equivalent to its being impossible to find a situation the occurrence of which verifies or falsifies the proposition.

32. *Meaningless Symbols, contd.* The fact that propositions may be strictly meaningless (except in the purely private and somewhat Pickwickian sense noted above) is of much more than merely academic interest

If we take it for granted that a proposition must be either true or false, then when we meet one which conflicts with conclusions to which other considerations have led us, we are obliged to set to work to demonstrate its falsity by one means or another (or perhaps seek for error in our own arguments, etc.). But if we can show that it is *meaningless*, i.e. that there is no conceivable means of verifying either it or its contradictory, then we can just toss it in the trash-bin and pass by on the other side. This will be found to save an immense amount of trouble, and to enable progress to be made in cases where otherwise nothing but inevitably endless disputation would ensue.

But it is very important to realize, if it is not already clear, the kind of proposition to which this treatment can legitimately be applied. This end may conveniently be promoted here by a few words about a curious essay in obfuscation recently attempted by a certain section of philosophers. This has taken the form of challenging the principle known as The Law of Excluded Middle—to wit 'A either is or is not B'; that is to say, 'between two contradictory statements there is no middle ground, both cannot be false, if one is denied the other must be affirmed' (*Webster*). In particular, it disputed whether a proposition 'must be either true or not-true (i.e. false)'. This 'major effort in bamboozlement', as Mr. Michael Innes says in a different and more entertaining context, adding, 'perhaps it was just the philosopher's instinct', appears to have been originated by Dr. Brouwer the Dutch mathematician and logician, and to have fluttered the metaphysical dovecotes to a quite unnecessary extent.[1]

[1] Bertrand Russell devotes several pages to it in his *Inquiry into Meaning and Truth*, for example. [*Inquiry into Meaning and Truth*, Ch. XX. (H. H. P.)]

The argument, as I understand it, runs somewhat as follows: It is futile to assert that a proposition is true unless we can prescribe means for verifying it; if this cannot be done, the assertion, like the proposition itself, is meaningless, but there are certain propositions, such as 'Snow fell in Manhattan Island on Christmas Day, A.D. 1', which it is impossible either to verify or to disprove; such a proposition accordingly cannot be said to be either true or false; but it is unquestionably a proposition, and therefore the law of excluded middle, which requires it to be one or the other (B or not-B, true or not-true) is unsound.

This, frankly, appears to me to be all rubbish, arising from a duplex confusion, (*a*) between the inherently impossible and the impracticably difficult; (*b*) between tautological certitude and the quantitative kind of assurance which is all we can ever attain in respect of any synthetic (matter of fact) proposition.[1]

To take the second first: The only sort of proposition about the truth of which we can be absolutely certain is the analytic or tautological proposition, of which the truth is assured by the definition of the symbols used. To take Ayer's example, the proposition '$2 \times 5 = 10$' is known to be absolutely true, simply because the symbols used are so defined as to make it so. No amount of empirical evidence could make us doubt it. 'It might easily happen . . . that when I came to count what I had taken to be five pairs of objects, I found that they amounted only to nine. . . . One would say that I was wrong in supposing that there were five pairs of objects to start with, or that one of the objects had been taken away while I was counting . . . or that I had counted wrongly. One would adopt as an explanation whatever empirical hypothesis fitted in best with the accredited facts. The one explanation which would in no circumstances be adopted is that ten is not always the product of two and five.'[2] And similarly in all such cases.

[1] Cf. Ayer on the 'strong' and 'weak' senses of 'verifiable' (*Language, Truth and Logic*, pp. 22–6).

[2] Ibid., pp. 96–7.

But in the case of synthetic propositions, dealing with matters of fact, this absolute certainty is never obtainable; there is always some chance that the relevant evidence is deceptive. If I assert that rain fell at Land's End on the morning of 2 October 1946, you will rightly judge the proposition to be probably—perhaps *almost* certainly—true, inasmuch as you have reason to suppose, or none to doubt, that I am a normally truthful person, and that rain often does fall at Land's End, etc. But I might, theoretically, be lying for some reason of my own, or have been misled by an unperceived aeroplane testing insecticide spraying gear, or have been hallucinated. The chance of my assertion being due to these causes may be small—perhaps not more than one in a million—but it can never be zero.

The proposition 'Snow fell in Manhattan Island on Christmas Day, A.D. 1' is clearly of exactly the same type. It is not inherently impossible to obtain (as Russell admits) evidence from geological and meteorological sources such as would enable one to form some sort of estimate as to whether the proposition is true. Admittedly we should have to say that our chance of being wrong is not very far from one in two, instead of the suggested one in a million in the case of rain at Land's End; but the difference is only one of degree, and that is not sufficient to justify putting the two propositions in different logical categories. To argue that, because it is not practicable to obtain in the one case data justifying a conclusion having so small a chance of being wrong as in the other, the proposition in the former case is neither true nor false—that snow neither did nor did not fall at the time and place in question—is merely childish, and a sad example of the unholy fascination that the gratuitous devising of pseudo-logical puzzles exerts over even the acutest intellects. The fact that we can never be quite sure which of the answers is right (as is true of every question concerning matters of fact) in no way refutes the contention that one or other must be.

The trouble has arisen in the first instance, I think, from confusion over the first point mentioned, namely the confusing of two uses of the word 'impossible', viz. 'inherently' and 'practicably' impossible. There is nothing *inherently* impossible about ascertaining whether snow fell in Manhattan Island on a particular day, even in the remote past—it just so happens that we can't do it with any high degree of assurance. But in principle it seems not only plausible but necessary to suppose that every day's weather conditions leave some sort of characteristic trace on the terrain concerned, so that a sufficiently accurate and informed examination would allow us to draw conclusions approximately as assured as those we form and accept on any other point.[1] This is an example of 'practical' impossibility, or extreme impracticability, not of 'inherent' impossibility. It is a perfectly good synthetic proposition, like any other, and not at all what I mean by a 'meaningless' symbol. There is no doubt whatever about the references or referents involved; it is only a question of whether the place in the physico-historical order claimed by the referents of the symbol is in fact filled by them or by those of some other symbol such as 'The sun shone all day . . . etc.'

On the other hand, there are plenty of statements which it is inherently impossible to verify, and of which it is correct to say that neither 'true' nor 'false' are applicable. In particular, all ostensibly synthetic (matter of fact) statements about alleged entities supposed *ex hypothesi* to be unobservable are of this type. If I say 'All Reality is One', having previously defined 'Reality' in some such terms as 'That which, itself unobservable, causes the Appearances which are all we can observe', there is no conceivable means of verifying the statement—if, by definition, you can't observe Reality, you can't tell whether it is One, or Many, or Two, or Twenty-nine-and-a-half.

But even here it would be stupid to say that such a sentence contravenes the law of excluded middle; one would merely say that it is not a proposition at all, but only an arrangement of words in grammatically propositional form; it is inherently

[1] The case happens to be an unfortunate one to discuss—it was not my selection—because snow produces very slight effects, and Manhattan happens to have been extensively built over, etc.; but this does not affect the principle involved.

impossible, *ex hypothesi*, to identify the referent or to place it. No referent can be found outside the dictionary: it is meaningless.

This *is*, whereas 'Snow fell in Manhattan Island . . . etc.' is *not*, an example of the type of ostensible proposition that I describe as 'meaningless'. No discussion of such pseudo-propositions can lead to anything but futile circularity, and all must be ruthlessly jettisoned.

A grammatically propositional form of words is a genuine proposition, is meaningful and worth discussing, only if it is *in principle*[1] verifiable by some sort of observation; that is to say, only if it (or some other proposition formally deducible from it) is such that certain observable results will follow if it is true and others, in principle observationally distinguishable from them, if it is false. Similarly, of course, for individual words: the word 'X' is meaningless unless the proposition 'X exists' (or *mutatis mutandis* in the case of non-substantival symbols) entails observable consequences such as will not be observable if the proposition is false, i.e. if X does not exist. We shall have occasion to apply this rule in important connexions very shortly.

33. *Expansion of Symbols.* I have already emphasized that discussion is waste of time unless we are sure what we are talking about. But we cannot be sure of this unless the symbols we use are subjected to a process which I shall call 'expansion'.[2]

In order that language shall serve its purpose as an instrument of communication, it is imperative that it should be, as Ogden and Richards put it, 'a *ready* instrument'; that is to say, it must in most practical situations be *concise*. The result is that we make very extensive use of what Humpty-Dumpty

[1] I repeat that *practicability* has nothing to do with it. The proposition 'The far side of the moon is red-hot' is meaningful (and possibly worth discussion by selenographers), though presumably false, because it is possible in principle to devise means of verifying it, e.g. by landing from a rocket-ship and burning one's toes, even if this be not a practicable enterprise at the moment.

[2] Some writers speak of 'reduction', and Ogden and Richards use the words 'expand',' expansion', etc., in a sense which seems to me somewhat different from mine, though I may have misunderstood them. But on the whole 'expansion' seems the least objectionable word here.

called 'portmanteau' words, containing, as it were, the maximum of information packed into the smallest possible space; and great confusion constantly arises from different people having, so to say, packed portmanteaus of indistinguishable appearance with very differing contents. The items included in Mr. Bevin's 'Democracy' portmanteau are by no means the same as those in that of Mr. Molotov—or even of Mr. Byrnes.

More particularly, for our present purpose, we constantly use words and phrases which on the face of them refer to facts of observation, whereas they actually refer to inferences from such facts, or to interpretations of sign-situations rather than to the signs themselves; and both the inferences and the interpretations may be erroneous.

For example: I may say to my wife, 'I saw Mrs. Jones on the Land's End road this morning'. She, perhaps having reason to suppose that Mrs. Jones went to London yesterday, may question this and ask, 'Are you sure it was Mrs. Jones that you saw?' To which I may reply, 'Well, I can't be *sure* —I wasn't near enough—but I certainly saw a red hat and a green dress; and as Mrs. Jones is the only person in the village, that I know of, who habitually wears such things, I concluded that it was she'. That is to say, my first statement is a contracted symbol, of which the second is an expansion. And better men than Mrs. Jones have probably been hanged on just such evidence as this.

But even my second statement is not nearly so fully expanded as it might be. If my sceptical consort presses me further, I should have to admit that I could not be sure that it was a *hat* and a *dress* that I saw, and to explain 'I saw a patch of red surmounting a patch of green, these patches being, so far as I could judge, of approximately the sizes and in approximately the relative positions which I am accustomed to associate with hats and dresses in such circumstances'—or words to that effect. And I might even be driven further and forced to expand the word 'saw' to some such statement as 'I enjoyed an experience of visual type, which I have no reason to suppose was hallucinatory'.

Note that at each stage I am doing precisely what we have described as *interpreting a sign-situation*. In the first, I have interpreted the signs 'red hat, green dress' as indicating the presence of Mrs. Jones; in the second I have interpreted the stimuli 'red patch, green patch' as signs of a hat and a dress; in the third I have interpreted the experience of visual type as a sign that normal seeing is going on.

Now, expansion of symbols under cross-examination in this kind of way might evidently be of great practical importance in certain circumstances, such as those of a criminal trial, or in studying the reports of a mediumistic séance, but it is equally clear that if we were to indulge it on every day-to-day occasions, communication would become altogether impracticable. That is why we have formed this inveterate habit of talking almost exclusively in highly contracted symbols, seldom even realizing that they are contracted, or that expansion to the full is necessary if we are ever to be sure what we are really talking about.

Usually we think we have attained this happy state when, for any doubtful symbol, we have substituted a set of other symbols from the dictionary. This is often a necessary stepping-stone. If, for example, we encounter the word 'martingale' we shall be wise not to jump to the conclusion that it is the name of some sort of bird (nightingale, house-martin), but to consult a dictionary and find that it is 'a strap . . . to prevent horses rearing, etc.', before committing ourselves to any expression of opinion.

That is good enough for ordinary conversational, etc., purposes; and if we are still in doubt we can fall back on ostensive demonstration: 'A *strap* is a thing like this or this or this; that and that are made of *leather*; there is a *horse*, and there's another', and so forth. But if we want to go further, and push to the limit the process of ascertaining what straps, horses, or other material objects 'really are'—to use a colloquialism which I think is harmless in the circumstances—then we must apply the same sort of expansion that we did in the case of Mrs. Jones, and examine the sign-situations on the interpretation of which our conclusions depend.

The statements 'There is a strap', 'There is a brick', 'There is a tomato', are of exactly the same type, not merely grammatically but logically, as the statement 'There is Mrs. Jones'[1] (and the same is true, of course, of such alternatives as 'Here is a brick', 'I see a tomato', etc., etc.) and must be expanded in the same kind of way.

When we have done so, and have decided what is being interpreted as a sign *of what*, and why, in circumstances occasioning statements of this kind, we shall find that we have said all that it is inherently possible to say about the Ultimate Nature of Matter. And when we have performed as closely analogous a process as possible in the case of mental objects (imaginings, dreams, hallucinations, etc., we shall be in a corresponding position with respect to Mind.

34. *Summary*. This concludes all that I propose to say here on the subject of Meaning. The account I have given has, I fear, been all too long for the reader's taste, yet all too short for adequate treatment of the topic; it will almost inevitably appear disconnected in some places, repetitive in others, and with far too many loose ends left undealt with. But I think it contains all that is necessary to our present salvation, and I will try to pull the essential points together into some semblance of coherence in the following short summary, before coming to grips at last with the main work of the book.

1. It is necessary to understand the Theory of Meaning, because (*a*) discussion is futile unless we ourselves know what we are talking about, i.e. unless we know the meanings of the words we use; (*b*) if we do not, we may be impressed or intimidated by the contentions or arguments of others, which superficially appear weighty or authoritative, and be misled into trying to refute them; whereas they may actually be quite meaningless, inherently incapable of either refutation or establishment, and therefore unworthy of discussion at all; (*c*) unless we properly understand the mechanism of

[1] [This is not strictly true. 'Mrs. Jones' is a proper name, applicable only to that particular person; whereas 'a strap' is a general description applicable to many different things. The author's point is, I think, that all those phrases are *material object* expressions. (H. H. P.)]

sign-situations and of the interpretation of signs, with which the Theory of Meaning is inextricably interwoven, we shall remain incapable, as philosophers have always been, of correctly analysing the simplest perceptual experience, which (as is both common sense and generally agreed) lies at the root of all inquiries of this nature.

2. There is no compulsory law which requires a word to have one meaning rather than another. That is to say, I use the word 'dog' when I wish to refer to certain common animals, rather than some other word, for reasons of broadly an historical character, not because there is any necessary or magical connexion between the word and the object. Subject to such historical, etc., factors, the use of words is entirely arbitrary and conventional.

3. This is as true of the word 'meaning' as it is of any other; and the word has been and is used in a great number of ways. Logically, we are entitled to earmark the word for any of these usages, adopting or inventing substitutes where necessary in other cases to avoid confusion. But it is only common sense to base our theory of 'meaning' on the process whereby words do, in fact, do their work as instruments of communication, and to define 'meaning' in corresponding terms. If we were to do anything else we should lose touch with the essential function of language, and become involved in side issues, such as its emotive or aesthetic aspects, which will lead us nowhere—as has usually happened.

4. A word does its work by virtue of being a *symbol*; and a symbol is a particular sort of *sign*, namely, that sort which is, in fact, used for purposes of thinking and communication.

5. A sign is a stimulus (or group of stimuli) which occurs as part of some larger group to which the stimulated organism reacts in some way or other, and which recurs as a part of some other group more or less similar to the first. The first stimulation and reaction modify the organism in some fashion; it is a fact of observation that, if part of a stimulus-situation recurs, then the organism tends to react in the same manner that it did to the original situation as a whole. This is usually expressed by saying that the first situation and reaction leave

'traces' of one kind or another (engrams, etc.) which are revived or re-excited, etc., by the recurrence of any part of that situation.

Thus, if the sound of the word 'Cat' is an element of a stimulus-situation containing, *inter alia*, the object Cat, recurrence of this sound as an element in another, broadly similar, situation will tend to make the organism (person, child, etc.) react in the same way as it did to the original situation—the term 'react' including the reviving, in so far as it in fact occurs, of the visual, tactile, etc., sensations initially experienced, in the form of images.

In more comprehensible if perhaps less accurate language, the sound 'Cat' becomes *associated* with the visual, tactile, etc., experiences initially presented in conjunction with it, and recalls these, as images, when it is re-presented.

This is what constitutes acting as a *sign*, and the psychological reaction to the sign is the organism's *interpretation* of it. 'Interpretation' is synonymous with the terms 'adaptation', 'state of mind', etc., and is known in Ogden and Richards's terminology as 'an act of reference' or 'a *reference*'.

To start with, the sound 'Cat' is a sign of the object Cat, but later it becomes a sign of a particular sort of reference, to wit a Cat-reference, in the speaker.

6. This reference *is* the 'meaning' of the word ('Cat') *for* the particular speaker or hearer *on* the particular occasion concerned. No other definition of 'meaning' can be satisfactory, because this is, manifestly, the way in which words do work and perform their communicatory functions.

7. It follows that no word, or other symbol, will ever mean quite the same for any two people, or for the same person on any two occasions, because the situations in which it has occurred will vary as between individuals, and the total relevant experience of any given person will change cumulatively, so to say, from one occasion to the next.

8. But communication is possible by virtue of the fact that, for any person and as between different persons, there will be a greater or lesser 'overlap' or 'hard core' of substantially identical experiences, and correspondingly of references. It

will be more successful as this overlap is greater, and less so as it is smaller.

9. The relation between the reference and the symbol, in speaker or hearer alike, is of the type commonly described as *causal*. A correct (successful) symbol is one which causes in the hearer a reference similar to that which caused its utterance by the speaker, with the proviso that the hearer must be a 'suitable interpreter', i.e. one whose relevant experiences have some considerable overlap with those of the speaker.

10. The 'object' (which may, of course, be either material or otherwise) to which the reference refers, or is 'directed', is known as the *referent*; and conundrums such as The Problem of Truth, etc., are most conveniently handled by speaking of different kinds of referents as being 'placed' in different 'orders', e.g. in the historical, physical, imaginary, dream, etc., 'order'. Thus a proposition (complex symbol) such as 'There is a stove in my room' may be correct and have a single meaning, inasmuch as it duly induces in suitable interpreters references substantially similar to that of the speaker; but these references may be false, inasmuch as they place the referent (stove in room) in the order of contemporary physical fact, whereas it belongs only to the order of (say) the speaker's memories or the like.

11. A true symbol (proposition) is one which not merely identifies or characterizes the referent (i.e. induces the proper reference) but allocates it to its proper order.

12. A symbol which has no referent is nonsense. In particular, a proposition is *meaningless* if it is inherently impossible (not merely very difficult) to make any observations tending to verify or refute it.

13. In practice it is usual and permissible (provided we do not lose sight of what we are doing) to short-circuit the detailed sequences of Referent referred to by the speaker, Symbol caused by his Reference, Reference induced by this symbol in the hearer, Referent of this reference, and to speak as if there were a direct relation between the symbol and the object to which it would commonly be said to refer. This object or referent, taken as substantially identical for

speaker and hearer alike, is then loosely but for most purposes adequately said to be 'the meaning' of the symbol.

14. In order to make quite sure what we are talking about, i.e. to identify our referents with complete precision, it is necessary first to proceed to the level of ostensive demonstration, and then to analyse the basic sign-situations involved.

symbol 'material object' refers, and we can only do this by studying what is actually happening in those situations which would commonly be said to justify such remarks as 'That is a (material) brick', 'Here is a (material) dog', or 'This is a (material) lemon.'[1] In other words, we must examine Perceptual situations, i.e. the general process of Perception, as most philosophers have agreed is necessary.[2]

36. *Perception* (1). Let us, then, take a typical perceptual situation, such as is commonly discussed in the books, and try to find out of what it consists and what is going on.

Allowing ourselves a slight change from our earlier diet of tomatoes, we will suppose that, on entering my kitchen, I enjoy the experience commonly known as seeing an object on the table and say (or think to myself), 'I see a lemon', and that, in the ordinary sense of the words, I am speaking the truth, i.e. that I am neither lying nor hallucinated nor the victim of a practical joke, but that *mirabile dictu* there actually is a lemon on the table. The situation is simple enough, but its proper logical analysis is distinctly tricky.

I do not wish to shock the reader unduly, but I think it important to point out at the start that, in such circumstances, the statement 'I see a lemon', though true (as we have

[1] Note that the word 'material', inserted in brackets above, is usually taken for granted and as 'understood' in ordinary communication, but this is only because we relatively seldom have occasion to talk about the bricks, dogs, lemons, etc., of dream, hallucination, etc., and when we do we say so specifically unless the context makes it plain. I shall adopt the same convention here.

[2] The standard work here is Professor Price's *Perception* (Methuen, London, 1932) to which, as to its author, I need hardly say that I owe much, though I shall not follow his treatment at all closely. Readers who desire to go more fully into the subject than I am able to do here, particularly as regards the difficulties that have been encountered in the past, and the alternative views developed by various schools of thought to meet them, should consult this work, also Professor Broad's *Scientific Thought* (Kegan Paul, London, 1923) and *The Mind and Its Place in Nature* (Kegan Paul, London, 1925), also Bertrand Russell's *The Analysis of Mind* (Allen & Unwin, London, 1921), *Our Knowledge of the External World* (first edition, Open Court Co., Chicago and London, 1914; second edition, Allen & Unwin, London), and *Mysticism and Logic* (Macmillan, London, fourth impression, 1921), not to mention many other writers, but particularly A. J. Ayer's two books, *Language, Truth and Logic* (Gollancz, London, 1938, second edition, 1947) and *The Foundations of Empirical Knowledge* (Macmillan, London, 1940).

supposed) 'in the ordinary sense of the words', is so only by virtue of the convention which allows us to employ highly compressed shorthand symbols whenever—as in 99 cases out of 100—it is impracticably cumbrous to use more fully expanded forms: it certainly is not strictly accurate, for it does not literally report observed facts at all, and short-circuits, so to say, a relatively elaborate act of what would ordinarily be called inference. Indeed, I doubt whether it would be too much to say that, in strict literality, the statement 'I see a lemon' is, in the circumstances described (or in any others for that matter), simply not true—unless, of course, we *define* the referent of the word 'lemon' by saying, circularly, that it is that which we actually see when we truthfully declare that we see a 'lemon'; but then we should need another and different definition of 'lemon' for the case in which we touch a lemon, or taste it, and so on. What I actually *see* in these circumstances is a yellow patch of a certain hue, brightness, shape, shading, etc., etc., against a brownish (table-top) background, and with various other visual experiences or data in certain relations to these.[1] And no one in his senses is going to say that a lemon *is* no more than a yellow patch of such and such a character.

Let us take this vitally important point from the opposite direction before we go any further. How do you define a lemon? Not, I hope, after all that I have said about Meaning, by using a dictionary and finding that a lemon is a citrous fruit and that a citrous fruit is a fruit of a lemon-like character. No. First we must decide whether we are defining the word 'lemon' or (as many people would say) the 'thing' lemon. We may here quote again with advantage from Ogden and Richards (*The Meaning of Meaning*, p. 110):

'. . . do we define things or words? To decide this point

[1] I should prefer to say, and in future usually shall say, that I 'cognize' or 'am conscious of' or 'am aware of' a yellow patch, because the word 'see' carries suggestions (in my terminology 'connotations') of optic nerve, retina, eye, light-waves, etc., which are in no way part of the immediate experience. Indeed, the use of the word 'see' virtually implies that the situation is one involving the presence of what we call a material object, whereas I might be hallucinated, in which case the word 'see' would not be applicable except by a certain perversion of its usual meaning.

we have only to notice that if we speak about defining words we refer to something very different from what is referred to, meant, by "defining things". When we define *words* we take another set of words which may be used with the same referent as the first, i.e. we substitute a symbol which will be better understood in a given situation. With *things*, on the other hand, no such substitution is involved. A so-called definition of a horse as opposed to the definition of the word 'horse', is a statement about it enumerating properties by means of which it may be compared with and distinguished from other things. There is thus no rivalry between 'verbal' and 'real' definitions.'

Thus the question is equivalent to asking what are the properties characterizing the referent of the word 'lemon' as ordinarily used; and this in turn to asking what observations and tests can be made to ascertain whether a particular object confronting us is of the kind referred to in such normal usage.[1]

37. *Perception, contd.* (2). We may say, then, that the word 'lemon' refers to, or stands for,[2] an object giving more or less the following results under test:

If you look at it, you see a yellow patch, which may be either of roughly elliptical or roughly circular shape, according to your position; if you take a knife[3] and cut it in certain ways you will find that it is enclosed by a thickish rind, of which the inner part is whitish and only a thin outer layer yellow; if you squeeze the rind in a particular way a strongly scented and flavoured oily substance exudes. If you pull the inside to pieces you will find that it consists of a number of segments of a more or less half-moon-wedge shape, each of which is enclosed in a fine skin, and consists of a fleshy sort

[1] Cf. Ayer on definition '*explicitly*' and '*in use*', in *Language, Truth and Logic*, pp. 66 ff.

[2] Note that we are, as usual, adopting the short-circuit view of 'meaning' here; strictly, we should say 'the word "lemon" symbolizes a reference to an object giving . . . etc.'

[3] It is my duty to point out that such words and phrases as 'look at', 'take', 'knife', 'cut', and any others that may be used, will themselves require definition—ultimately, if need be, by ostensive demonstration, as 'Knife—thing like this'; 'take, cut—do *thus* and so.'

of substance further describable on demand; each such segment contains, as a rule, one or more objects known as 'pips'. Pressure on these segments produces a fluid which has an acid taste, and the performance of certain further operations on this enables us to obtain the observations known as identifying the presence of the substances 'citric acid' and 'vitamin C'. One could continue the list almost indefinitely.

It does not matter here how many of these tests are necessary or sufficient to identify any given object as a lemon, for culinary or scientific or other purposes, or in how many it could fail to conform to specification and still be counted as one. Obviously, if its colour were a few shades more inclined towards the red, if it tasted sweet instead of sour, and were approximately spherical instead of elongated, the plain man at any rate would be disposed to say that it was not a lemon but an orange, i.e. to use the word 'orange' and not the word 'lemon' in referring to it. But the botanist might insist on continuing to call it a lemon, on the ground (say) of its having the number of chromosomes characteristic of lemons in general but not of oranges;[1] but in this case he will be defining the word 'lemon' as referring to a fruit of a certain (citrous) type having a particular number of chromosomes, which may, of course, be more important from his point of view than such trivialities as sweet or acid taste, which principally interest the plain man.

The point is that it is only by *some* procedure such as this that we can possibly ascertain whether the object is or is not a lemon, i.e. whether it does or does not conform to the specification which constitutes 'being a lemon' for whatever purpose, or from whatever point of view, is in fact adopted. The fact that no more than one or two easily ascertainable properties (results of tests)—notably colour and shape—are usually sufficient for all practical purposes has nothing to do with the logic of the matter. Note that, as a matter of fact, colour and shape alone, though almost universally relied

[1] I have no notion whether the number differs; very likely not, but this does not affect the illustration.

upon when buying lemons—and usually safely so—are peculiarly fallible for purposes of definition; for they equally well appertain to the waxen imitations sometimes used for decorating sideboards, or to those pieces of toilet soap supplied by the ingenious Mr. Culpeper in happier days.

It is obvious, I think, that when we make any statement about a lemon, the word 'lemon' must be and always is taken to be a shorthand symbol for 'an object which gives such-and-such results to such-and-such tests'. If not (that is to say, if the object to which the word 'lemon' is in fact used to refer fails to do so in any important respects) then we are not talking about a lemon but about something else. And to 'see' all that is thus compressed into the shorthand symbol 'lemon', when all that is actually cognized is a yellow patch of a certain shape, etc., would be extending the notion of vision even beyond the range of Sam Weller's 'million magnifying power microscopes'.

38. *Perception, contd.* (3). How then are we justified in saying 'I see a lemon'? The short answer, good enough for most purposes, is 'By experience'; that is to say, previous experience of similar yellow patches in similar contexts leads us to expect, maybe with very considerable assurance, that *if* we do thus and so, we shall have further experiences of this and that. But this is not certainty, and the only thing of which we are absolutely and irrefutably certain is the awareness or cognition of the yellow patch.[1]

[1] As Professor Price says (*Perception*, p. 3), 'When I see a tomato there is much that I can doubt. I can doubt whether it is a tomato that I am seeing, and not a cleverly painted piece of wax. I can doubt whether there is any material thing there at all. Perhaps what I took for a tomato was really a reflection; perhaps I am even the victim of some hallucination. One thing, however, I cannot doubt: that there exists a red patch of a round and somewhat bulgy shape, standing out from a background of other colour-patches, and having a certain visual depth. . . . What the red patch is, whether a substance, or a state of a substance, or an event, whether it is physical or psychical or neither, are questions that we may doubt about. But that something is red and round then and there I cannot doubt. Whether the something persists . . . whether other minds can be conscious of it as well as I, may be doubted. But that it now *exists*, and that *I* am conscious of it—by me at least who am conscious of it this cannot possibly be doubted.'

The better answer and one more in conformity with the general line we have been taking, is to say with Ogden and Richards[1] that the perceptual situation is essentially a Sign situation. This should indeed already have occurred to the reader, for the very mention of expectations based on past experience indicates a state of affairs, past and present, precisely identical with that discussed in my brief account of Signs generally in section 25 above. The yellow patch has in the past been one of a larger group of stimuli—roughly but sufficiently describable as experiences of lemons—which have included the various tactile, visual, olfactory, gustatory, etc., sensations corresponding to the processes of squeezing, cutting, dissecting, etc., mentioned above. To these stimulus groups or situations, I have reacted in one way and another, and the reactions have left physiological or psychical 'traces', or both. When I again encounter the yellow patch stimulus these traces are revived, and constitute my psychological reaction to it, as determined by these previous experiences and by the present situation. The yellow patch, in short, is a *sign* to me of those other properties (observations, results of tests) which together characterize a lemon and differentiate it from other sorts of objects; in other words, I interpret the yellow patch as being a sign of all these properties, and, if it actually is a normal, material lemon that I see, then in due course some or all of them will be observed—in the words of Ogden and Richards, the 'external context' will have been 'completed', and my 'act of reference' will have been true. But if some important group of these properties be absent, then my expectation is disappointed and my reference is false. My reference, in these circumstances, will have been to a normal, material lemon; if, on attempting to cut it, it suddenly explodes in my face, I conclude that it is not a lemon as usually defined, but a bomb planted for my undoing by some infuriated metaphysician or, if the tactile properties are wholly absent, that I am the victim of a visual hallucination.

39. *Perception, contd.* (4). We must now consider in

[1] *The Meaning of Meaning*, Ch. IV.

greater detail the processes whereby the truth of the reference is in practice verified, whether deliberately or otherwise or, perhaps preferably, expand to the limit the statements by which they would ordinarily be described.

In the ordinary way, of course, we do not feel it necessary to take steps to verify the reference at all; we simply react to the situation without thinking and the verification occurs automatically in due course, and there is no question of doubt or belief, as some writers seem to think. But just as I would conclude that I was hallucinated if, on attempting to pick up the lemon, my fingers closed on empty air, so, if for some reason I suspected hallucination, this is the sort of test I should apply. To assure myself that the indubitable visual experience (yellow patch) was not hallucinatory, I should presumably go through the motions of stretching out my hand and attempting to grasp the putative lemon. If on doing so my fingers encountered a familiar type and degree of resistance, I should conclude that I was dealing with at least some kind of a material object (though not necessarily a lemon as defined); whereas, if they apparently passed through it, and met without any sensation of resistance being experienced, I should come to the opposite conclusion. The first inference is not necessarily correct, because the experience of resistance (sensations of touch and pressure, etc.) might itself be hallucinatory; but hallucinations of two senses at once are so extremely rare that the possibility could safely be neglected for all practical purposes.

Materiality then, is sufficiently assured by visual plus tactile observation, assuming each to be appropriate to the other under the conditions obtaining at the time and place concerned. The reservation is necessary, because experience of material objects in general has shown us that there is a very strong correlation between certain types of tactile sensation and certain types of visual sensation; so that, if the object *looked* like a prolate spheroid with slightly pointed ends as a lemon does, but *felt* like a rectangular box with sharp corners, we should conclude that there was something very wrong indeed somewhere.[1] It does not, of course, matter whether the visual or the tactile observation comes first; the principle would apply just as well if I were to find the object by touch in the dark and then switch the light on. But it will save trouble if we adopt the convention of supposing that the visual observation is first in the field, and say that the non-hallucinatory character of an object is assured if the visual observation (cognition of the yellow patch) is followed by the appropriate tactile observations (cognition of touches, pressures, etc.) under certain conditions.

40. *Perception, contd.* (5). The last three words are important, because, in general, there is a gap to be filled between the visual experience and the tactile (or whatever other may constitute observation of the result of any test). This is represented above by the words 'the motions of stretching out my hand and attempting to grasp', and the words 'stretch' and 'grasp' must be expanded, if our story is to be complete.

What is the full (or sufficiently full) expansion of the sentence 'I stretch out my hand'? We will leave the words 'I' and 'my' for future consideration, while the word 'hand' must obviously be dealt with on the same lines as the word 'lemon'. The question here concerns the phrase 'stretch out'. I do not think there is any great difficulty about the answer. When I stretch out my hand (more accurately, perhaps, my arm, but no matter) I am aware of certain sensations (i.e. as I shall put it later, I cognize certain cognita) which, to speak a shade colloquially, are located in or originate from my elbow and shoulder joints and the muscles of my arm. I do not normally *attend* to these with any particularity, but this is of no importance here; anyone who wishes to quibble on the ground that he is 'unconscious' of these sensations may conveniently be asked how, in that case he knows (eyes shut) whether it is an arm or a leg that he is stretching. Physiologically speaking, these sensations are produced by the stimulation of certain receptors in my joints and muscles, and the

[1] I shall deal below with the case of sticks in water, etc., which are crooked to the eye but straight to the touch.

transmission of nervous impulses to my brain, just as my sensations of sight and touch are produced by the stimulation of other sorts of receptors.[1]

Concurrently with these I cognize a dull pinkish patch of a certain shape (the visual aspect, so to say, of my hand) which moves in a certain way; that is to say, in order to keep the position of this patch in my field of vision unchanged, I have to turn my head or move my eyes in their sockets, and these movements, like those of my arm and hand, are themselves represented in consciousness (if I may be allowed this phrase for a moment), in so far as I attend to them, by certain other kinaesthetic sensations proceeding from the muscles of my eyeballs and neck, etc.

I only mention the eye movements, etc., for the sake of completeness; the essential part of the story is the awareness of the kinaesthetic sensations from my elbow and shoulder, arm muscles, etc. We can, as we all know, perfectly well stretch out an arm in the dark, though not, as a rule, with great accuracy, and there is nothing unnatural about this; but it is a very curious experience to *see* one's arm moving, if one cannot *feel* it moving, i.e. if, for any reason, the kinaesthetic mechanisms are out of action; and no amount of eyeball swivelling or hand-gazing will of themselves make the hand move.

Precisely similar remarks apply, of course, to the finger movements of 'grasping', which I accordingly need not further consider here.

It should be noted that these sensations of endosomatic origin enjoy exactly the same status, so to say, logically and physiologically, as those of exosomatic origin. There is no respect in which we are required to give the one class priority over the other. The kinaesthetic, etc., sensations may, it is true, usually be less closely attended to, or less vivid, etc.,

[1] Sensations originating from within the body are said to be of 'endosomatic' origin, and include all those starting from the viscera, etc., as well as those starting from joints, muscles, etc. The latter, when relevant to awareness of movement, are termed 'kinaesthetic'. Sensations originating from outside the body, as by the incidence of light, sound, heat, etc., or pressures, etc., are said to be of 'exosomatic' origin.

than the visual or tactile sensations arriving from outside the body; but both are equally part and parcel of the reference-checking procedure.

41. *Perception, contd.* (6). The next point I want to make is important. Not only are these kinaesthetic, etc., sensations part and parcel of the process of verifying or reference—i.e. of making sure that we are dealing with a normal, material lemon, and not a visual hallucination, a plaster cast, or a booby-trap bomb—in the sense that they are a necessary part of it;[1] they must occur in the proper *order* with respect to the other, exosomatic, sensations which we call the results of tests, or observations, made 'on' the (putative) lemon. We should not consider that we were dealing with a normal lemon if we became aware of tactile sensations, however appropriate, when our hand was resting on our knee and had not been outstretched; or if, the moment we grasped the knife, and before any cutting had been attempted, the peel suddenly flew off and the segments fell apart in a litter, so to say, of little 'pigs'; or if we found our mouth full of pips without any antecedent biting, etc. I do not know quite what we should conclude in such circumstances, but certainly that the situation was highly abnormal, as regards either the lemon or ourselves or both. It does not matter, as remarked above, whether the visual or the tactile sensation comes first, but there must always be an alternation of what I may loosely call 'operational' with 'observational' sensations, cognizings or awarenesses; and it is imperative that these should follow each other in proper sequence.

It will presumably be conceded without further ado that what is true of the simple test of stretching and grasping applies also to any other type of test. Whenever we speak of any operation, such as squeezing or slicing or peeling, or weighing or measuring, or chemically analysing by processes of heating, mixing, adding reagents, etc., we are invariably speaking of what would ordinarily be called objects and

[1] I trust I need not explain in detail here that precisely similar considerations, with but slight alteration, would apply to other more elaborate or indirect methods of verification, as by prodding it with a stick or causing someone else to pick it up, etc.

movements, of one kind or another. In order to be sure what we are actually talking about, these objects and movements must, in principle, be analysed and expanded in precisely the same way that we have done for the lemon and the stretch-grasp test above. This is the only way in which we can be sure that the object before us does in fact conform to the defining specification of whatever it is supposed to be—by which, in other words, the referent of the name given to it can be placed in its order, failing which, any remark made must be meaningless.

I shall return to this point later, but I wish now to make a short digression on a matter of terminology.

42. *A Point of Terminology.* In the foregoing account of the simple test of stretching and grasping, I have used the word 'sensation'; but I have done so solely in deference to the reader's (particularly the lay reader's) probable habits, and not because I think it the best word to use. On the contrary, it has certain very grave disadvantages for the purposes of a discussion of this kind. To say, for example, that I have a sensation of yellow inevitably carries suggestions of light waves of a certain frequency falling on the eye, being focused on the retina, there causing chemical changes in certain substances, and initiating impulses which pass up the optic nerve to certain areas of the brain. There is no objection to this in the ordinary straightforward case in which we are in fact seeing a normal, material lemon, etc., in the ordinary physical sense of the term—for this is how 'seeing' in that sense is defined.[1] But these suggestions or implications are quite out of place if the putative (material) lemon is in fact hallucinatory; for it is, in effect, part of the defining specification of an hallucinatory object that there are *no* light waves falling on the eye, *no* impulses passing up the optic nerve, etc. And the same applies, if possible more strongly, to the vivid imagining of a lemon or to dreaming of it.

On the other hand, there is no conceivable justification for

[1] It is important, however, as I shall emphasize below, to remember that any statements about eyes, nerves, etc., including statements that they exist, must themselves be amenable to the same kind of treatment we have given to the lemon. They equally refer to material objects.

saying that the entity (yellow patch) which, as we should say, we 'see', or are aware of, in the case of hallucination is in any way intrinsically different from the one we are aware of in the case of a material lemon. If the two were distinguishable by inspection, so to say, we should not be hallucinated; for the essence of hallucination is that the data (yellow patch, etc.) of which we are actually aware are so like those of which we would be aware if a material object were present that we mistake the one for the other with consequent misinterpretation. And there is no conceivable test which we can apply to the intrinsic character of these entities. The tests we can and do apply bear on the nature of the object to which, so to put it temporarily, they belong; and it would be a quite illogical begging of the question to say that two yellow patches *must* be intrinsically different, *because* one belongs to a material lemon and the other to an hallucinatory lemon.

Similar objections apply to such terms as 'sensa', 'sense-data' or 'sense-content', which are those most used by philosophers at the present time when they wish to speak of the immediate 'objects of awareness' in perceptual situations —and too often in other connexions also; and they apply with even greater force to the habit of saying that we 'sense' sensa, etc. The point is that the entity of which we are immediately aware inevitably comes first, and the question of whether it is, as we would ordinarily say, of sensory origin always has to be settled later by means of the kind of test we have been considering.

We clearly need a word by which to refer generically to all these entities or existents of which we are immediately aware, or are immediately 'given in consciousness', without prejudice as to their material (sensory), hallucinatory, oneiric or imaginary origin or (perhaps preferably) relations. For this purpose I propose to use the words *cognitum* and *cognita*, and to speak of cognizing a cognitum, where most philosophers would speak of sensing a sensum (or sense-datum or sense-content). And I shall speak of *cognizables* if I wish to refer to these entities when not in the state or relationship known as being cognized; that is to say, a cognized cognizable is a cognitum and a cognitum is a cognized cognizable. Anything whatsoever of which I am immediately aware, or is an immediate object of consciousness, or part of a field of consciousness, or mentioned in any like phrase, is a cognitum; and anyone who says that he knows, perceives, is conscious or aware of, or cognizes anything other than cognita is, by definition, guilty of a contradiction in terms.

The word *cognitum* accordingly refers to a class of existents (actually the only class of existents, but we need not bother about that here), including the referents of all such words as 'sensa', 'sense-data', 'sense-contents', 'percepts', 'images', 'ideas', 'impressions', 'sensations', 'feelings', 'appearances', etc., etc.

43. *Cognita, contd.* The foregoing definition is, of course, purely tautological—and rightly so. If it were anything else it would necessarily make some assertion, express or implied, either about the 'nature' of cognita or about their provenance, and such assertions would require verification, which would involve making observations, i.e. the cognizing of cognita. With one doubtfully possible and highly recondite reservation, which I shall mention later, nothing can be said, so far as I can see, about the ultimate nature, constitution, etc., of cognita; they are irreducible.

This is not to say that what we might well speak of as a single cognitum may not be analysable into constitutive elements. It is perfectly legitimate to speak of cognizing a yellow patch of a certain shape, and it is convenient to speak of this as 'a cognitum'; but there are obvious differences between cognizing an elliptical yellow patch and cognizing a square one, or between either of these and an undifferentiated field of yellow indefinitely extended. In the first instances we cognize, so to put it, ellipticity or squareness, as the case may be, as well as yellowness; in the third, only yellowness. Whether it would be correct to say that what is cognized in the first case is a yellowness cognitum plus an ellipticity cognitum (plus, of course, the cognitum or cognita forming the background), or whether we should say that all possible

[1] Cf. p. 122, below.

elliptical patches of any particular shade of yellow form a kind of family of cognita on their own, and whether, if so, the members of that family are discrete or continuous with each other—these, and many others like them, are questions well calculated to give healthy exercise to psychologists and logicians, but they do not concern us here. All that matters, and it is vitally important at that, is to realize clearly that any conceivable experience, whether of material objects, of hallucination, of events in dream, of imagination, or of the ecstasies of mystical experience, *must*, in the last analysis, consist in the cognition of cognita of some kind; and that, correspondingly, any statement about any experience must finally reduce to statements about the nature[1] and relationships of the cognita cognized. Any attempt to contend otherwise is equivalent to saying that the experience concerned consists in being conscious of nothing, and this (*pace* Heidegger[2]) is meaningless non-sense.

Possibly I ought to devote a few lines here to guarding myself against those who might object that three kinds of mental state or process have been traditionally recognized by psychologists, namely Cognitive, Affective, and Conative (non-technically equivalent respectively to Knowing, Feeling, and Striving), and that all this talk about cognition of cognita refers only to the first of these. This is not warranted. Properly speaking, all states are cognitive, though it may be difficult enough to analyse them, to locate the origin of the cognita cognized (the phrase is somewhat colloquial but will serve here), or even to give a satisfactory name to the state experienced at a particular moment. It is not easy, and may even be impossible, to introspect fully on the memory of a state of anger, for example (and I think certainly impossible while it is affecting one), or to sort out and identify the obscure complex of endosomatic sensations (sensa, cognita)

[1] By 'nature' here, I do not mean 'what they are made of', or equivalent foolishness, but whether they are what we refer to by such words as 'yellow', 'red', 'salt', 'sweet', 'loud', 'high-pitched', etc.

[2] [Professor Martin Heidegger, the author of *Sein und Zeit*, etc., one of the founders of the Existentialist philosophy. His doctrine concerning *das Nichts* ('Nothing') is a favourite *bête noire* of Logical Positivist writers. (H. H. P.)]

which constitute it; and it may be far from easy to decide whether it is or was predominantly one of anger or of fear. But the mere fact that such states, however difficult of analysis, are none the less distinguishable sufficiently guarantees their cognitive character for our purposes.

44. *Perception, contd.* (7). Let us turn back for a moment to our lemon. We have seen that the process commonly known as Perception is essentially a process of interpreting a Sign-situation. It is also essentially expectant, anticipatory, or predictive. When, cognizing a yellow cognitum of a certain shape, I say that I see a lemon—the adjective 'material' being understood—I am using a condensed shorthand symbol of which the full (or nearly full) expansion would run somewhat as follows: 'I cognize a yellow cognitum of such-and-such a kind (elliptical, pointed at the ends), in a setting or context of certain other cognita; I have cognized substantially similar cognita in sufficiently similar contexts before; these previous cognitions formed parts of sequences consisting of, or importantly including, groups of cognita of the arm-stretching and finger-closing ('grasping') type, followed by certain other cognita of tactile, pressure, thermal, etc., types (corresponding to the observations that material lemons are more or less smooth, firm, cool, etc.); and so on and so forth for all the other observations which make up the defining specification of a normal material lemon; these groups of cognita of various types, cognized in a certain sequence or pattern, are my interpretation (i.e. psychological reaction or reference to) of the yellow patch, which is a sign of them to me.'

'This is all very well', the reader may say at this point, 'but you have been talking all this time about the meanings of words and what any ordinary person would call the *properties of* the lemon; what I want to know about is the *lemon itself* which *has* these properties and is what *I* mean by a material object made of *matter*.'

The short answer is, of course, in the immortal words of Sairy Gamp, quoted at the head of this chapter, '*Wich I don't believe there ain't no Mrs. 'Arris*'. That is to say, there is no such thing as 'the lemon-itself', or any other 'thing-in-itself' or *Ding an sich*; and any statement about such an alleged entity is, at best, completely meaningless. The lemon, or any other material object, simply *is* the sum of its observable properties—or, better, just *is* its properties (or qualities, attributes, etc., to linguistic taste)—and to speak of an observable property is to speak of a cognitum sequence or pattern.

If you insist that there *must* be a lemon-itself, even if no one has as yet been clever enough to find it, then I am, of course, open to conviction. You are in effect asserting the proposition 'There exists a lemon-itself' (let us call it a *Lem* for short), or 'A *Lem* exists'. How do you propose to verify this statement? What is your referent, and how do you propose to 'place' it?[1] In particular, can you deduce any observable consequence from the alleged existence of the *Lem*, which would not be observable if the *Lem* did not exist; or can you give a defining specification of the *Lem*? If you can, well and good, but in that case the verifying observations, which are necessarily cognitum sequences of one kind or another, merely get added to the schedule of lemon-cognita; but if you cannot, then it is inherently impossible either to verify or disprove your statement, which is accordingly meaningless, since it has no identifiable referent. However closely and thoroughly you may examine, analyse, or study the lemon, you will never find anything except (if you are lucky) new properties (cognitum sequences), and the fact that it is linguistically convenient to speak of 'the properties *of* "the lemon"' adds nothing to the facts of the situation. The lemon-itself, like every other *Ding an sich* (*Ding* for short hereafter), remains, and *ex hypothesi* must remain, for ever and inherently unobservable; and this is equivalent to its not existing at all—at least there seems to be no other rational meaning to be attached to the phrase 'not exist'.

Your only possible way out, it seems to me, is to take some such line as those who hold the so-called Causal Theory of Perception. You might not implausibly, on the face of it, argue that there must be some *reason* why the cognitum

[1] On the 'placing' of referents, cf. pp. 69–70, above.

sequences known as the observable properties of any material object (which I maintain *are* that object) form the characteristic pattern they do, and not some other pattern, or a mere haphazard configuration; and that this is equivalent to saying that there must be *something* which *makes* them take that pattern. But this does not help. Regardless of what you may mean, or suppose yourself to mean, by the word 'reason' (or 'cause', if we adopt slightly different wording) the referent of the symbol is either inherently observable or not. If it is, then any observations on it will consist of cognita, as before; if not, then the symbol is meaningless. Statements involving the word 'cause', etc., are in fact no more than convenient ways of talking about certain observable regularities in phenomena (observations, cognitions of cognita), notably of antecedence and subsequence; and the notion of *necessity*—probably derived from primitive experience of pushing things about—was adequately disposed of by Hume,[1] many years ago. But, as Russell says, '. . . we want to be able to *feel* a connexion between cause and effect, and to be able to imagine the cause as "operating". This makes us unwilling to regard causal laws as *merely* uniformities of sequence; yet that is all that science has to offer.'[2] And so we try to smuggle in the thing-itself, implausibly disguised as an operating cause.[3]

The truth of the matter is that there is not the smallest justification for talking about the thing-itself, the *Ding*. It is (though even this is a contradiction in terms) purely a piece of 'gratuitous metaphysics', as Russell puts it in substantially

[1] *Treatise of Human Nature*, Bk. I, Pt. III, Sect. 14, 'Of the Idea of Necessary Connection'.

[2] *The Analysis of Mind*, p. 89.

[3] If anyone wishes to say that cognitum patterns, sequences, configurations, etc., are 'caused' by 'forces' acting between them, in the same sort of sense that we say that the movements of planets or electrified bodies are caused by forces acting between them, then I have no objection; indeed, I shall say something very much to this effect at a later stage myself. But it is important, in this case, to remember that to talk or calculate in terms of 'forces' is merely to employ a symbolic device for rendering observations amenable to treatment; we never, properly speaking, observe forces, we only observe the planets, etc., together with clocks, scales, and balances, i.e. cognize certain groups and sequences of cognita.

this connexion[1]—it is a mythological monstrosity. It does not exist.

45. *The Metaphysical Ding.* Yet the tacit or explicit assumption that there is this *Ding* behind or beyond or within the properties, qualities, attributes, appearances, which we are commonly said to observe, has poisoned the springs of philosophic thought from time immemorial, and still poisons them; so that even so advanced a thinker as Professor Price feels it legitimate to speak[2] of 'physical occupants' of space, coincident with the 'families of sense-data' (i.e. groups, etc., of cognita, in my language) and serving, in Ayer's words, solely as 'the inferred subjects of as causal characteristics'. But as Ayer points out, in effect, all that we do when we speak of causal characteristics is to enunciate a hypothetical proposition about sense-data (cognita).[3] To say that event B is caused by event A, is only to say (omitting certain elaborations) that the cognitum pattern known as event B is regularly preceded by the cognitum pattern known as event A (or words to this effect). All we can know are cognita and groupings, sequences, etc., thereof; and if we want to talk about causation at all, we must frame our remarks in conformity with this fact—i.e. in terms of regularity of succession of cognitum patterns—and abstain from dragging in inherently unknowable non-existents, all reference to which is necessarily meaningless.

I want here to make it clear that it is not, as might be supposed, permissible to claim that the existence of the *Ding* can be *inferred* from observations of the qualities, etc., it is said to 'have' or the 'effects' it is alleged to 'produce'. Many people, I suspect, imagine that such an inference would be

[1] 'If we can state the laws' (regularities) 'according to which the colour' (of a fading wallpaper) 'varies, we can state all that is empirically verifiable; the assumption that there is a constant entity, the wallpaper, which "has" these various colours at various times, is a piece of gratuitous metaphysics. We may, if we like, *define* the wallpaper as the series of its aspects', i.e. of the 'appearances' or cognitum groups, etc., which would commonly be said to be appearances 'of' the paper. *Our Knowledge of the External World*, first edition, pp. 106–7.

[2] *Perception* (Methuen, London, 1932), Ch. IX.

[3] *The Foundations of Empirical Knowledge*, p. 227.

logically parallel to inferring, for example, the existence of the planet Neptune from observations made on the planet Uranus. But this is not the case. Inferences of this latter type are essentially hypothetical in the sense that they can be tested, and retained as facts or discarded, according to the outcome of the test. In the case of Neptune, the test consists in turning a telescope towards a particular point in the night sky and seeing whether a small bright disk is or is not observable at the predicted place and time; or, alternatively, if Neptune happened to be black, by seeing whether the observed perturbations of Uranus continued to be such as would be caused on the assumption that there was a planet of the estimated mass at the calculated distance, etc., from the sun. Note importantly, however, that even if Neptune were black and therefore invisible, this would not mean that it was inherently unobservable; it might, for example, be possible to locate it by radar, or to verify its presence by its passage across the face of a more distant star, etc. But the *Ding*, on the other hand, *is* inherently unobservable, for no conceivable experiment could possibly reveal anything but cognitum groups, etc., which are exactly what it is invented to account for.

Note also that the *Ding* has not even, as I once erroneously supposed, the modest status of a Fiction. The essence of a fiction is that something happens *as if* some proposition were true which in fact is not—as we might treat the property of an incurable madman *as if* he were dead. But this is only possible if we know what would happen if he really were dead—and we do; it would be senseless if we had no experience of dealing with the property of actually deceased persons. But to say that the cognita constituting our so-called 'knowledge of' a material object hang together, or follow each other, or are patterned *as if* there were a *Ding* responsible for the patterning, is to beg the whole question at issue. It would be rational only if we could say with assurance what would happen in the two cases of there being and not being a co-ordinating *Ding*.

Finally, the fact that to speak on a thing-itself is meaningless, and precisely equivalent to speaking of nothing at all, does not necessarily imply that to do so is a mere harmless eccentricity, which anyone is free to indulge without risk if he feels so disposed. At first sight it might appear that no more harm is likely to be done by using verbal forms which assert or imply the factual existence of the *Ding* than if we were to make a kind of pious ritual of writing 'Z' on both sides of every equation, or ending every sentence with the exclamation 'Selah!' And this is true, provided always that we know and remember precisely what we are doing, namely, using such signs as ornaments, or emphatics, or (in the case of the 'thing') merely as a linguistic convenience. The trouble is that we forget this, and unthinkingly get into the habit of taking it for granted that such words as 'thing' or 'object' have referents which exist and must somehow be taken into account when we are considering or discussing material phenomena—as we shall see in a moment when we come to examine the principal views about Perception taken by philosophers at the present time.

46. *Current 'Theories' of Perception.* I do not propose to devote much space to considering the opinions, all equally erroneous in my judgement, advanced and discussed by philosophers concerning our knowledge of material objects; but it will be worth while running through the more important of them in at least a cursory way.

Nearly all philosophers agree, I think, that there is some sense in which the entities I call 'cognita' do exist and are 'real'—though some, I understand, are so perverse as to admit their existence while denying their 'reality'. This is incomprehensible to me, unless they use the word 'real', as do many of the laity, as synonymous with 'solid' or 'tangible'—in which case they would do better to say so—but we will let that pass for the moment. But there is the greatest diversity of opinion as to the view we should take regarding material objects—the pens, tables, chairs, lemons, etc., of our familiar environment, and about the sense in which we may be said to know them.

Curiously enough, but perhaps characteristically, the only point on which all schools seem agreed, is the erroneous one that we do not know any object *directly*, but only 'through' or by virtue of what are nowadays most usually called 'sensa' or 'sense-data', which form, of course, according to me, a class of the entities I have called 'cognita'.[1] As a matter of convenience I shall use the words 'sense-datum' and 'sensum' indiscriminately myself in the immediately following passages; but it is to be understood that by doing so I mean no more than cognita so related to other cognita that they would ordinarily be said to be of sensory origin, or words to that effect.

The introduction of these terms has resulted from the criticism, based on what is known as The Argument from Illusion, of the now moribund school of Naïve Realists, who held that when (as we say) we see or touch a material object we are actually, in the plain everyday sense of the words, directly and immediately perceiving or aware of that object itself—and that, in effect, there is no more to be said.[2]

The line on which this simple view has been attacked is summarized by Ayer[3] as follows: 'The argument, as it is ordinarily stated, is based on the fact that material things may present different appearances to different observers, or to the same observer in different conditions. . . . For instance, it is remarked that a coin which looks circular from one point of view may look elliptical from another; or that a stick which normally appears straight looks bent when it is seen in water. . . . The familiar cases of mirror images, and double vision, and complete hallucinations . . . provide further examples. . . .

[1] The words 'sensum', 'sensa', are completely synonymous with 'sense-datum' and 'sense-data', usage being entirely a matter of individual preference.

As already indicated, many other words have been used by various writers; for example, and not guaranteeing the allocations, which are from memory—'ideas' (Berkeley), 'impressions' (Hume), 'sensations' (Mach), 'percepts', 'appearances', 'particulars' (Russell at various times—his use of 'sensibilia' is somewhat different, as I understand it). 'Sense-data' has also been used by Russell, and 'sensa' by Stout and Broad.

[2] The wording here could probably be improved upon, but it will sufficiently serve the present purpose. Note, however, that this view is a good deal nearer the facts of the case than its critics would allow, though not quite in the sense its proponents supposed.

[3] *The Foundations of Empirical Knowledge*, p. 3.

The same thing occurs in the domains of the other senses, including the sense of touch. . . .'

Since, it is argued, a coin cannot be both circular and elliptical, while our moving into another position cannot reasonably be held to alter the shape of the coin (which, incidentally, may be reported unchanged by another observer who does not move), it is not easy to say just what we mean when we say that we see the coin, unless we say that what we are immediately aware of is not 'the coin itself' but something else which is in some way related to it—to wit a 'sensum', 'sense-datum', etc., according to taste.[1]

'These sense-data,' says Ayer (ibid., p. 2) 'are said to have the "presentative function"[2] of making us conscious of material things. But how they perform this function, and what is their relation to the material things which they present, are questions about which there is much dispute. There is dispute also about the properties of sense-data, apart from their relation to material things; whether, for example, they can appear to have qualities they do not really have, or have qualities they do not appear to have; whether they are in any sense 'within' the percipient's mind or brain.

No wonder. If you insist on gratuitously postulating a non-existent *Ding* ('the coin', etc.) it is not surprising that you have difficulty in deciding how anything is related to it. Knock out the *Ding*, and the difficulty disappears; we have then only to consider how 'its' (so-called) appearances (cognita) are related to each other, and that is merely a matter —logically simple, if sometimes practically elaborate—of writing down a strictly factual account of what actually occurs. The dispute here is about a pseudo-problem which should never have been allowed to arise. As for asking whether sense-data can have properties they do not appear to have, or vice versa, this is not even a pseudo-problem; it is sheer contradictious rubbish. 'Can an appearance appear other than it appears?' 'Can a red sensum (cognitum) appear

[1] I cannot forbear to draw the reader's attention to the way in which trouble arises the moment this villain of the piece—'the coin', 'the coin itself', etc.—appears on the scene.

[2] Price, loc. cit., p. 104.

green?' and like foolishness. Such questions are preposterous. A red cognitum which appears green is a contradiction in terms; a straight cognitum-group which appears crooked is another—and so on. Any green appearance *is* a green cognitum (or group, etc., of course), and any crooked appearance *is* a crooked cognitum. This second class of dispute arises from a kind of reduplicated gratuitousness. We start by postulating an unobservable *thing* which *has* 'appearances', and then go on by asking whether these appearances themselves may have 'qualities' (i.e. appearances) which they do not appear to have—and thus launch ourselves, so far as I can see, on a potentially endless regression of argument about the appearances of the appearances of the appearances of . . . etc.

However, this kind of thing is really not worth powder and shot, and we will pass on to a brief account of the current theories at present more or less in vogue.

47. *Current 'Theories' of Perception, contd.* 'There are three main ways in which philosophers have attempted to modify the Naïve Realist thesis that visual and tactual sense-data are parts of the surfaces of material objects, so as to make it defensible. . . . We may call them respectively the Theory of *Multiple Location*, the Theory of *Compound Things*, and the Theory of *Appearing*.'[1] Each of them attempts to answer the Argument from Illusion 'by saying that in illusion the distorted or dissociated or otherwise errant sense-datum *is* part of the surface of an object but "with a difference"'.[1] Note that the all postulate the existence of 'an object' *to* the surface *of* which the sense-data are in some way related.

The Theory of Multiple Location (due wholly or mainly to Professor Whitehead) 'says that we must distinguish between the characteristics which characterize something only *from a place*, and those which characterize it *simpliciter*. From a particular place the penny's top surface really is elliptical and smaller that the top surface of a sixpence: but simply—*in itself*[2] or from no place—it is circular and twice as large. . . . The penny just *is* elliptical from this and that place . . . exactly

[1] Price, loc. cit., p. 55. [2] My italics, W. W. C.

as in itself it is circular and smooth in outline.' (Note the words 'in itself' again.) A thing, '. . . in so far as it is simply located, i.e. in so far as Science is concerned with it . . . would not be red or green or hard. But in so far as it is multiply located, it really will have these qualities.'[1] A more baffling way of saying what we know already, namely, that a penny looks round from one place and oval from another, without adding anything to this knowledge, would, I feel, be difficult to devise. Frankly, I do not at all know what it all means; and I venture to doubt whether anyone else does either. To say that a *thing* can *be* circular *from* one place but elliptical *from* another place seems to me to be as nearly senseless as anything short of actual gibberish can well be. But, if we leave 'the penny' (i.e. the *Ding*) out of it, we can report what actually happens in a rational manner, by saying something to the effect that there is a non-chance relation between the (visual) cognita which constitute what we call the (visual) surface of the penny and those (visual and otherwise) which constitute being in a certain place, or moving from one place to another.

The Theory of Compound Things says (I quote from Price again) that 'the illusory visual and tactual sense-data . . . do really form parts of the surface of an object, but of a *compound* object.[2] The stick is not bent, but the compound stick-plus-water really is bent, and the crooked sense-datum is part of its surface. These compound objects really do exist in external Nature and do have their qualities (which differ from those of their constituents taken singly) just as "simple" objects like sticks have theirs.' This is, perhaps, not quite so bewildering as the story about Multiple Location; but the notion that two inherently unobservable entities—the 'stick-itself' and the 'water-itself'—somehow combine to form a third—to wit the 'stick-plus-water-itself'—seems to me but little if any less preposterous. Incidentally, what combines with the penny-itself to make the new compound object,

[1] Quotations from Price, loc. cit.

[2] This theory appears to be due to Professor Alexander. [S. Alexander, *Space, Time and Deity* (Macmillan, London, 1920), Bk. III, Ch. VII; cf. also Ch. VIII. (H. H. P.)]

penny-plus-something-itself which now has the quality of ellipticity instead of roundness? The distance-itself, the space-itself, or the position-itself? And echo answers What? Again it is easy enough, if only we are content not to drag in all these 'objects' (*Dinge*) to give a straightforward account of what happens, in terms of observations made on the visual shape of the stick, its tactile shape, and its immersion in water, etc.; and these observations can, in principle, be fully expanded into terms of cognitum sequences.

The Theory of Appearing (at one time, at least, held by Professor Prichard and Professor Moore[1]) adds nothing, so far as I can make out, to the ordinary colloquial account, in which we say that the penny 'looks' oval, the stick 'looks' bent, etc. '. . . the theory holds that a visual or tactual sense-datum is always part of an object's surface appearing to someone to have certain characteristics. On one form of the theory "appearing" is the name of a unique and unanalysable *three-term-relation* between part of an object's surface, a characteristic or set of characteristics, and a certain mind. On another form of it, "A appears *b* to Smith" stands for a unique and unanalysable kind of *fact* about A, *b*-ness and Smith's mind, but this fact is not of a relational sort.' (Price, ibid., p. 62.) I cannot see that this sort of thing does more than 'make it more difficult', like the snake in the story, and Professor Price is right in saying of this theory that 'its own foundations are incoherent'.

48. *Current 'Theories' of Perception: Criticism.* These so-called theories are, of course, open to criticism, even if taken at their face value, and (I suppose) are capable of some kind of defence, on what I call purely technical grounds; and they have, in fact, been destructively criticized by Professor Price and others, whose work should be consulted by those interested. But it would be waste of time even to summarize this criticism here; for to meet the proponents of such theories on their own ground, and discuss whether their views are true

[1] H. A. Prichard, *Kant's Theory of Knowledge* (Clarendon Press, 1909), Ch. IV; G. E. Moore, *Philosophical Studies* (Kegan Paul, London, 1922), pp. 245–6. (H. H. P.)

or false, is not the right way of dealing with them. They are neither true nor false, but meaningless, one and all; and criticism of meaningless propositions must itself be meaningless or else wholly irrelevant and therefore no criticism.

Ayer goes a long way in this direction, though not quite far enough, when he contends[1] that these 'theories' are not really theories at all, but only alternative languages. 'Can we', he asks, 'discover any empirical evidence that favours one of these theories rather than another . . .?' and goes on to point out that we cannot, 'For we find that every conceivable experience, in the field to which these theories refer, can equally well be subsumed under any of them. Each of them will cover any known fact; but none of them, on the other hand, enables us to make any inference at all from the known to the unknown. No matter which of them we adopted, we should be able to describe our perceptions, whatever their nature; what we should not be able to do would be to make any predictions. But if the relation of these three theories to the relevant phenomena is precisely the same, then, as theories, they are not distinguishable from one another. And if they allow no possibility of extrapolation, if the actual course of our experience can have no bearing upon their truth or falsehood, it is misleading to call them theories at all.'

This is perfectly sound so far as it goes, and if these theories were all dealing with some one set of existents, then I think it would be correct to describe them as 'alternative languages', i.e. as merely different ways of saying the same thing, or different symbols standing for the same referents. But this is not the case, for they all purport to be concerned with, and to make statements about, the relations of sense-data (cognita) to 'an object' or 'the surface of an object'; and these terms, as I have insisted over and over again, are symbols of which the referent cannot conceivably be identified, and are therefore meaningless. Thus the equivalence of the three theories is not due to their all saying the same thing

[1] *The Foundations of Empirical Knowledge*, pp. 53 ff.

in different words, but to their all saying *nothing* in different words.

Let the reader cast back to even the few and condensed quotations I have given (or, better, consult Professor Price's book, or the original enunciations of these views), and he will see how 'the object', 'the object's surface', 'the object in itself', etc., permeate the whole discussion. Delete it, and the whole fabric crashes to the ground. It is nothing short of astonishing to me that men of the outstanding intellectual calibre of Whitehead, Alexander, and Moore, not to mention many others, should have failed so much as to notice, let alone challenge most vehemently and at pistol point, this enormous nigger at the very centre of the verbal woodpile they were trying to sort out, but have taken it for granted that there must be an 'object' which 'has' properties, or is in some way responsible for the sense-data they cognize. Tyrannous indeed has been the bondage of verbal forms.

Incidentally, my view that it is not quite correct to describe these three alleged theories as alternative languages, is, I think, borne out by the reflection that, if they were, it should be possible not merely to translate from one into the other, but to give rules for doing so; and I doubt whether this would be possible. But provided we realize clearly that they none of them mean anything at all, the point is of minor importance.

The Causal Theory, on which I have already touched above, is, of course, in no better case than these; for, as Ayer justly remarks,[1] 'The essential point to bear in mind is that, in every case, the object that is singled out as the cause of what is immediately observed is not itself supposed to be observable'. And to talk about one sort of inherent unobservable is no more meaningful that to talk about another sort. To make remarks involving the use of the word 'cause', etc., is, of course, a perfectly legitimate thing to do, provided we are clear as to what we are referring to, namely, certain types of observable sequence of events (cognitum sequences); but to insist that there 'must be' an object' or 'thing' which *causes* the cognita or their patterning is as meaningless as to talk of

[1] Loc. cit., p. 173.

a thing-itself which *has* properties, etc. There is no conceivable reason, to put it colloquially, why one cognitum group or sequence should not be the 'cause' of another, i.e. stand to it in that relation of regular precedence, etc., which we describe as 'causal'. To invoke a mysterious unobservable to do the work is much as if we insisted that the celestial bodies must slide in grooves cut in an unobservable 'firmament' or the like, on the ground that otherwise they 'could not' follow the determinate paths observed. We do not do this, but say that the paths are determined by 'forces' acting between the bodies. It is true that we cannot observe the forces, but, on the other hand, we do not allege that they are existents in the sense that 'the object itself' is supposed to be. The word 'force' is, I think one would say, an auxiliary symbol enabling us to handle our actual observations in a useful sort of way. It actually refers, I think, to some rather complex symbol (set of propositions, etc.), itself referring to certain cognitum sequences described as some material body, a scale and a clock, and coincidences between elements in these sequences—but I would not claim that this is a very accurate statement. In any event, the use of the word 'force' does not involve the invoking of some inherent unobservable; our raw material consists strictly of observations, and the referent of force, whatever it may be, is constructed out of these.

Similarly, when we find cognita behaving in a highly systematized, regular, and determinate manner—notably going about, so to say, in those gangs or 'families' (Professor Price's term) which we describe as 'material objects', it is absurd to chatter about their relation to a mythological and mystical[1] Object, *Ding*, or What-not, which *must* be there somewhere in order to hold them together—for all the world as if the 'properties of a thing' were hotel labels stuck on a suitcase. We should be better advised to follow the lead of the physicists, and see whether we cannot work out, *from our observations*, the 'laws of force', or equivalent, which must be supposed to act between them so as to make them behave as

[1] I use the word here in a purely pejorative sense.

they do—with due reference to their possible dissociation or aberrancy, or their interaction with other systems.[1]

49. *Preferable Views of Perception.* If we abjure all this nonsense about things-in-themselves, etc., we find that, on what I may term the practical level, all difficulties about bent sticks, elliptical pennies, doubled candles, etc., are very easily dealt with by going almost to our starting-point, and regarding the perceptual situation as what in practice it actually is, namely, a matter of the interpretation of signs. I need not recapitulate all that I said about signs, for the gist of the matter is simple enough.

Experience teaches us that a circular brownish, etc., patch, in the context of the situation known as 'looking perpendicularly at' (or equivalent) is a sign of the tactile experiences known as 'feeling circular' to come in due course; but it has also taught us that 'oval patch' in the context of 'looking obliquely' is also a sign of circular tactile sensations to come (the appropriate movements of stretching and grasping, etc., being assumed in each case). Again, the group of cognita described as 'straight visual stick' (or 'sight of straight stick') normally presages 'straight tactile stick' sensations, and 'crooked visual stick cognita' presage 'crooked tactile stick cognita'; but if the 'crooked visual stick cognita' are encountered in a context which includes 'stick immersion in water cognita', then they are a sign, by virtue of previous experience of 'straight (or less crooked) tactile stick cognita'. Similarly, two visual candle cognitum groups are normally a sign of two tactile candle groups; but if they occur in a context including eye-pressing cognita they are or may be a sign of only one tactile candle.

Expansion into this kind of form leads to awkwardness of wording, as the reader will have remarked; but this is only because the language does not lend itself to such usage, and I think my point will be sufficiently clear. In accordance with the basic theory of signs, our interpretation of any particular

[1] A hint of the way in which W. W. C. would probably have explained the phenomenon of 'Psycho-kinesis'. Cf. also App. III below, pp. 250–3, where this 'possible dissociation or aberrancy' is used as the explanation of Precognition. (H. H. P.)

group of stimuli (or, more generally,[1] of cognita) depends on the contexts in which we have encountered them in the past, and on our present situation; and we shall expect the relevant tactile experiences to be round or elliptical, straight or crooked, double or single, according to whether or not the context in which they occur includes the stimuli (cognita) corresponding to perpendicular vision, water-immersion, or eye-pressure, as the case may be.

But if we want something that might not implausibly be called a New Theory of Perception, let us forget all this nonsense about things-in-themselves—which means nothing—and appearances which appear other than they appear—which means, if possible, less—and try the effect of a little radical positivism. Let us be bloody, bold, and resolute, and adopt the desperate expedient of saying firmly what is manifestly and unchallengeably true. The visual penny *is* elliptical (except in special cases), while the tactile penny is circular; the visual stick *is* bent (look at it!), while the tactile stick is straight (feel it!); there are two *visual* candles though only one *tactile* candle; and so forth.[2]

I am not sure whether this is worth giving a name to—though better so than its predecessors, perhaps—or whether it can properly be called a Theory, or what name to give it; but I should think that something like the Doctrine of Autonomous Appearances would serve, provided the word 'Autonomous' is not understood as excluding correlation. There is, of course, a very close correlation indeed (we may say 'non-random relationship' if the purely mathematical flavour of 'correlation' be deemed objectionable) between visual and tactile cognitum groups (not to mention gustatory, olfactory, etc.), so long as we are dealing with what we call material objects; indeed, as I have already insisted, it is only by

[1] According to the author's theory one should say 'more accurately' rather than 'more generally', since the term *stimulus* would require a very great deal of 'expansion'; though it is the appropriate one to use in the quasi-Behaviouristic theory of Ogden and Richards. Cf. pp. 55–60, above. (H. H. P.)

[2] Cf. Russell on double vision. 'If we see two tables, then there really *are* two visual tables.' *Our Knowledge of the External World*, first edition, p. 86. (H. H. P.)

observing these relationships that we know whether we are dealing with material objects or not, and only by means of them that material objects can be defined. But this is not because they are tethered to a *Ding* like goats to a peg; in principle they are free to come and go 'like the blessed gods' at their own sweet will, or rather under what we would ordinarily describe as the influence of whatever forces act on them—which, actually, comes to the same thing if their comings and goings are non-random.[1] The point is that, on this view, there is nothing in principle to prevent (say) the visual components of a material object, so to call them, existing and being cognized independently of the tactile and other components, as there is if we regard them as being forcibly co-ordinated by the 'thing-itself' of which they are commonly said to be appearances.

This enables us to cover the phenomena of hallucination, dream, imagining, etc., much more comfortably than can be done by the pseudo-theories discussed above. In fact, we are able to deduce at once the possibility of encountering a group of visual cognita, so related as to form a visual object, but unaccompanied by the tactile and other groups which, duly co-ordinated with it, would form the 'family' constituting a material object, though it would not follow that we ever *must* do so. But, as a matter of fact, we sometimes do, and then we call the experience an hallucination or a dream or an imagining, according to whether we are awake or on the edge of sleep, and to how vivid the visual cognita are, etc.

I am inclined to think that this entitles the doctrine to rank as a true theory, though I have no desire to press its claims. If, for reasons of bashfulness or otherwise, no one had ever reported an hallucination, for example, we might plausibly argue, I think, that any phenomenon not positively excluded from possibility is likely to happen sometimes, and would predict that sufficiently careful search ought to reveal instances of the cognition of visual objects lacking their normal tactile concomitants (or vice versa, etc.), and adequate examination of witnesses would verify our prediction. Thus, in principle,

[1] Cf. Eddington, *The Nature of the Physical World*, (Cambridge University Press, 1927), pp. 147 ff.

our view would enable us to make inferences from the known to the unknown, and to test the inferences empirically; and this is the characteristic of a genuine theory. However, the point is of no importance here.

50. *What Matter Is.* It should now be plain enough what sort of answer we must give to the question What *is* Matter?

As a rather quibbling formality to clear the ground we start by saying 'Matter is a symbol[1] used to refer to the whole class of material objects', and we go on to say that a material object *is* a sequence of cognita or cognitum-groups related in a particular way, or of which the relationships conform to a particular pattern.

And if we are asked *what* particular way or particular pattern we are referring to, we reply that it is that or those which are recognized and studied by (material) physicists;[2] that is to say, those in which the sequence of visual, tactile, auditory, thermal, gustatory, olfactory, etc., and not omitting the kinaesthetic (visceral, intra-muscular, arthritic, etc.), cognitum groups follow each other in conformity with those regularities known as the laws of (material) physics.

In short, a material object is a certain kind of cognitum sequence—neither less nor more—and if you want a more exact specification you must, in principle, apply to the physicists, who will tell you what relations must hold between the different types of cognita for the sequence to qualify as being of this 'certain kind'.

Strictly speaking, there is no more to be said, logically, in reply to the question 'What is Matter?', but there are one or two points I should like to note in passing.

[1] Cf. what we said about Truth on p. 73, above. If anyone objects, and declares that 'Matter' should refer to the 'stuff' or 'substance' out of which things themselves are made, the answer is that you obviously can't make a non-existent *Ding* out of an existent 'substance', and that all that we have said about things-themselves applies equally to 'substance'.

[2] I insert the qualifying word 'material' here because I shall later contend that physics probably can be, and, if so, certainly should be extended to deal with the whole range of what we commonly call 'mental', 'psychical', and 'spiritual' objects and phenomena, as well as the 'material', and to subsume these just as it has subsumed Chemistry and is in process of subsuming Physiology and Biology in general. [Cf. Russell, *Analysis of Mind*, pp. 305 ff. (H. H. P.)]

First, I do not much care for the plan sometimes adopted of saying that a material object *is* a 'logical construction' *out of* sense-contents (cognita),[1] or words to this effect. Cognita I know, and their flux and sequence and change I know and these *are* the material and other objects I deal with, according to their patterning. But to say that my table *is* a logical construction out of these seems to me to be moving too far away from cognizable reality, and it is not at all clear to me what the identifiable referent of the symbol 'logical construction' is.[2] But I may be wrong here, and the point is a small one.

Second, there is the hideous question of whether cognita are truly irreducible. For all and every purpose of this work there can, of course, be no doubt at all that they are, for nothing that could fairly be called study or analysis of any object, material, or otherwise, or of any cognitum or cognitum group, could possibly yield anything but cognita. Yet I am not quite so happy about it as I might be, and certain questions gnaw at my mind. Is it conceivable that by applying a sufficiently generalized mathematics to sufficiently (i.e. completely) generalized raw material, we might in principle be able to deduce the whole of physical and non-physical phenomena as logically necessary and complete consequences[3] —somewhat as Eddington claims that all field physics, and the constants of Nature, can be deduced from relata and relations and some relation of likeness between some of the relations, plus a restriction to a fourfold handling, and the principle of Indistinguishability? If not, how are we to answer the question of why things happen as they do and not otherwise? But, if so, then would we not be, in effect, getting, as it were, behind cognita and so 'reducing' them—or would we? I do not

[1] e.g. by Ayer, *Language, Truth and Logic*, pp. 73–4, 189–90.

[2] Professor Ayer gives a clear account of the meaning of this phrase in *Language, Truth and Logic*, p. 73. (H. H. P.)

[3] Note that this would not constitute *apriorism* in the usual and objectionable sense. If, for example, our axioms were only such as assert that entities exist and are related, there would be no question but that they were empirically verifiable. Cf. Eddington, *Nature of the Physical World*, Ch. XI.

[[4] What meaning could be given to this question on the author's theory? Cf. pp. 116–17, above. (H. H. P.)]

think, however, that we need worry about this here, though to me at least it is a not unattractive line of speculation.

Third, it is important to realize that—often despite appearances—physicists are also human and in no better case than lesser men as regards the entities of which they can have direct knowledge or awareness. They no more than we can cognize anything but cognita. It is true that they concentrate on configurations of cognita somewhat different from those that chiefly interest the likes of us—notably on those coincidences of light and dark patches known as readings on scales, balances, and clocks, or those black and coloured ones forming what we call spectra, etc. But a red patch is no less a cognitum and no more so when it is part of a spectrum than when it is part of a sunset. It is perfectly right, proper, and legitimate for a physicist to say, as a matter of convenient verbal shorthand, that he is studying the path of an electron, when what he is actually doing is cognizing certain elongated white patches (lines) on the dark background of a photographic print, just as it is legitimate for us, on the same grounds, to say that we are observing a lemon when we are actually cognizing a yellow patch on the brown background of a table. But it would be as futile for him as for us to contend that anything is actually going on, other than the cognition of a cognitum group, and the interpretation of it as a sign of other cognitum groups and sequences to be as expected.

Unfortunately, physical scientists no less than others, and perhaps even more, have been brought up, tacitly and by implication, even if not explicitly, in the old tradition of classical metaphysics—of substance and attribute, and 'things' *having* 'properties', etc.—and with no regard at all to Meaning, the theory of signs, the functions of language, or to the pitfalls hidden in the use of shorthand terms. As a result they have become, oddly enough, the most devout of metaphysical myth-worshippers, insisting with almost fanatical zeal on the 'reality' of Matter,[1] supposedly in some way conditioning the

[1] It cannot be more than some fifty years ago, I think—which is but yesterday in the history of human ignorance—that the late Lord Kelvin (an eminent physicist in his day) was declaring that the one thing of which scientists were absolutely certain was 'the substantial reality of the aether'.

observations which they make 'on it', and quite ignoring the 'reality' of the mind or consciousness by virtue of which alone the observations are made.

It is true that the notion of substance has virtually faded out of modern physics, as it was pretty well bound to do when sub-atomic physicists found it necessary to describe electrons, etc., not merely in terms of 'waves' (which might have been compatible with some sort of 'substantial' aether), but of 'waves of probability', or something very like this.[1] But the antiquated superstitions about 'matter' and 'substance' and 'thing-themselves' die hard, and it cannot be too strongly insisted that all that any physical scientist has ever cognized, regardless of what he may say about it, or can cognize or ever will be able to cognize, are cognitum groups and sequences.

51. *Reality and Existence.* At this point someone is almost sure to object that by declaring material objects to be 'only' or 'no more than' (as they will tendentiously insert) cognitum sequences of a certain kind, I am thereby denying the 'reality' of matter, or its 'existence', or else am maintaining that it is 'really' mental, or is an illusion.

Such objections would be entirely incorrect. I have not said a word to imply that cognita are mental, let alone illusory; and there can be no doubt at all about the reality or existence of the material objects amid which we live and move and have our being. All that I have been concerned to deny is that the word 'matter' refers to anything at all beyond 'the whole class of material objects', or some closely equivalent expression, and that there exist inherently unobservable 'things' or 'substances' other than cognita we cognize. But such errors are pardonable enough in view of the havoc that the words 'real', 'really', 'reality', etc.—with 'exist' and 'existence' running them close—have wrought in philosophic thought from one generation to another. It will accordingly be worth while to devote a few paragraphs to the discussion of these words.

Well over two thousand years ago Parmenides (with whom, according to Hegel, philosophizing proper began) wrote a philosophical poem, of which the first part was called 'the

[1] Cf. Lindemann, *Quantum Theory*, Ch. III.

way of truth' (Reality), and the second 'the way of opinion' (Appearance); and only the other day, as things go, Mr. Bradley gave to the world his well-known metaphysical essay entitled *Appearance and Reality*;[1] and throughout history philosophers, and more amateur thinkers also, have been obsessed by the idea that somewhere behind or beyond or above or within the world of 'appearances' lies some magical, mystical, imperceptible world of 'reality', of which the appearances are at best but a shadow or a symbol (in the non-technical sense, if any) and between which and us hangs a veil of ignorance through which our limited intelligence can no more than feebly penetrate. It is my categorical affirmation that the whole of this is complete and utter balderdash and arrant nonsense.

More meekly I submit that the words 'real', 'reality', etc., never add anything except emphasis to any sentence in which they occur, and that they should be expunged once and for all from the vocabulary of serious discourse, except in such limited and purely conventional usages as 'real and virtual images' (optics), 'real and imaginary numbers' (mathematics), or 'real and personal property' (law). The word 'real', I maintain, is precisely equivalent to 'conforming to definition' (or 'defining specification')—neither more nor less. As such it is always redundant, since an alleged 'X' which does not conform to the defining specification of an X is simply not an X—for example, an object judged to be a lemon by sight alone[2] might turn out, on further investigation, to be made of plaster of Paris; we should then, properly speaking, say simply that it is *not a lemon*, as ordinarily understood, but not that it is 'unreal'. In practice we might very likely say that it was 'not a real lemon'—implying that it was an imitation or artificial one; but this is only because the words 'natural' and 'material' are commonly taken as understood in all ordinary circumstances; and these words are equivalent to 'conforming

[1] F. H. Bradley, *Appearance and Reality* (first edition, Swan & Sonnenschein, London, 1893; second edition, 1897. Reissued by the Clarendon Press, Oxford, 1930).

[2] i.e. an object such that certain visual cognita were interpreted as signs of other lemonoid cognitum sequences to be expected.

to the schedule of observable properties (cognitum sequences) which it is agreed define a natural and material lemon'.

The statement 'Queen Anne is really dead' adds nothing except emphasis to the statement 'Queen Anne is dead', or 'Queen Anne is not really dead' to 'Queen Anne is not dead'; and the question, 'Did you really spit in Hitler's eye?' adds nothing, except perhaps a flavour of incredulity, to the question, 'Did you spit in Hitler's eye?'

If I meet a friend who yesterday was clean-shaven, and observe him to be proudly sporting a noble beard, I may colloquially remark 'I don't believe that's a real beard'. By this I do not mean that the beard is unreal in the sense of non-existent, and I am unlikely to mean that it is a pure visual hallucination, or a too vivid image of my own. I almost certainly mean that I do not believe it is a *beard* (natural and material) as ordinarily understood and defined, to wit an assemblage of bristles growing out of follicles in the skin, but rather an imitation of a beard so defined, made by fixing hairs to a bit of muslin and attaching this to the face with spirit gum. Whether this or some similar hypothesis is correct, or whether my friend has sufferent a miraculous sprouting in the night, is a matter which can be tested by experiment.

If *no* process of observation or experiment will reveal any difference between the allegedly 'real' (conforming to specification) object and the allegedly 'non-real' (artificial, imitation, putatively illusory, etc.), then we have no justification for distinguishing between them so far as what would ordinarily be called their nature or constitution is concerned. So-called 'cultured' pearls, for example, are produced, I understand, in precisely the same way as are so-called 'real' pearls, namely, by the deposition of calcareous slime by an oyster around some irritant particle of grit; but in the one case the presence of the particle is accidental, while in the other it is deliberately inserted by the ingenious cultivator. If the other conditions are the same, then no chemical analysis or physical test (presumably) will distinguish the 'real' from the cultured variety, and the distinction is purely a matter of history; it is between 'accidentally induced' and 'deliberately induced', and not between 'real' and 'unreal'.

52. *Reality and Existence, contd.* (2). Instances of this kind could be multiplied indefinitely, but the reader will probably complain that this is not at all the kind of thing he has in mind when he uses the word 'real' or, more particularly, the word 'unreal' or some equivalent phrase. He will say that the antithesis is not between 'real' and 'artificial' or 'imitation'—for he understands perfectly well that plaster lemons or artificial pearls are just as 'real' existents in the world as are their natural counterparts—but between 'real' and 'imaginary' or 'hallucinatory' or the like. Surely, he will ask, you do not contend that the miraged water of the desert is as real as that in which I bathe, the Centaurs of my dreams as real as the horse I ride next morning, the pink rats of my alcoholic delirium as those I catch in my traps?

If by 'real' he means 'material' (notably 'solid', or 'liquid', as the case may be), then, of course, I do not; but in that case he should *say* 'material' (or 'solid' or 'liquid'), and not 'real'. The mirage is a perfectly real mirage, inasmuch as it conforms to the specification of a mirage as (roughly) water you can see but not wet your feet in; it cannot possibly be anything else;[1] the dream Centaur consists of undeniably real images, i.e. entities having the defining properties of images but not of material objects; and the hallucinatory rat is an undeniably real hallucination, i.e. a set or sequence of visual cognita not correlated with or followed by the tactile, auditory, olfactory, etc., cognitum groups (operational groups of kinaesthetic, etc., cognita being assumed duly included in their proper places) which make up the specification of a zoological rat. In other words, these objects *are* visual mirages, centaurs and rats (the word 'real' being, as always, redundant), but they are not all-senses-in-proper-sequence objects, so to put it; that is to say, they *are not* water-pools, hippanthropoids, or rats, as these objects would (or might) be defined by physicists or biologists.

[1] To speak of an 'unreal mirage' could only mean that the object in question was not a mirage.

I repeat, the word 'real' can always be replaced by some such phrase as 'conforming to specification'—whether of material object, image, hallucination, or anything else—and *ipso facto* at once becomes redundant, because, if the object we are talking about does not conform to the specification of the referent of the word we use, then we have used the word wrongly and are talking about something other than we suppose.

Every existent is 'real' *in its own order*—material, hallucinatory, oneiric, imaginary, etc., etc.—and to say that X is unreal is only another and misleading way of saying that we have misinterpreted a sign, or are allocating an existent to an order other than its own. Outside of the technical usages mentioned above, the word 'unreal' can properly be applied, if at all, only to non-existents (though this, of course, is almost a contradiction in terms). One might reasonably use the word, for example, in speaking of 'square circles', which certainly do not and (by definition) cannot exist even in imagination; but not, I think, of 'glass mountains', which may exist easily enough in the order of images or dream.

Let us by all means use the word 'real' and its accomplices colloquially, as heretofore, for the sake of emphasis or added intelligibility, calling a spade a *real* spade if we wish, or remarking that Lord X was *really* a cheap and nasty vote-catching scoundrel; but let us abjure the error of supposing that the word adds anything whatever to the communicative content of the sentence.

53. *Reality and Existence, contd.* (3). The word 'Reality' is, if possible, a more serious source of confusion than the word 'real'. We have not space here, of course, to examine or even to mention the innumerable varieties of nonsense that have been talked by philosophers in this connexion; but I think it is worth while, as so often, to note in passing the perfectly natural psychological causes that have given rise to it.

Philosophers, like the rest of us, have often been deceived by their fellow men. They have bought their gold bricks, and found them brass, or fortune-making oil stock of value only as pipe-lights; and have sadly reflected 'He seemed to be an ıonest man, but was *really* a common swindler'. Such common experiences as these, which are in fact no more than misinterpretations of signs and misallocations of referents to orders, have sharpened the contrast between the so-called apparent' (beneficent stranger) and the 'real' (dishonest trickster), while a lack of understanding of the theory of signs has obscured what has actually been going on.

More generally, philosophers, being for the most part sensitive and intelligent men, have been revolted by the strange blind callousness of Nature, in which kindness and mercy and the grace of humour seem to have no place; and above all by the gratuitous cruelty, violence, treachery, and sheer nit-witted imbecility of human beings, who know so much better and do so much worse than the fires and tempests and the beasts of the field.[1]

What more natural than to argue that, if appearances may be deceptive in one regard, they may be so also in another; so that, just as the beneficent-seeming behaviour of the oil-salesman was only the deceptive appearance of a noxious reality, so all the manifest evil and malevolence of Nature and of Man may be only an Appearance of a Reality which, if not necessarily Good and Beautiful and True, will at least be sufficiently different from superficial experience to be tolerable to the enlightened mind. And, of course, the step from 'may be' to 'must be' is psychologically a very small one.

To the elucidation of the nature of this Reality philosophers have extensively devoted themselves for many centuries, but, not unnaturally—since every possible observation is condemned and rejected *ex hypothesi* as a 'mere appearance'—have 'evermore come out of the same door as in they went', to wit, the covers of the dictionary.

Probably the physicists, when they try to philosophize—or

[1] It is all-important to remember, however, if sanity is to be preserved, that, although the more extreme forms of malignancy appear to be the sole prerogative of man—for it is difficult plausibly to attribute any great moral obliquity to earthquakes, or even wolves or puff-adders—yet it is from human behaviour exclusively that we derive our notions of such higher virtues as those just mentioned. Ghosts of dead cynics, please note.

some of them at least—are the worst and most dangerous of all offenders in this sort of connexion. A particularly pernicious abuse of the word 'real', for example, is to be noted whenever they tell us, as they delight in doing, that something or other is not 'really' what it obviously is to common sense, but quite different. The lay mind is, to be sure, obsessed by the idea that nothing is 'real' unless it is solid—or at least solidifiable—but the physicist takes a perverse pleasure in assuring us that nothing is 'really' solid at all. Thus Eddington declares, 'My scientific table is mostly emptiness. Sparsely scattered in that emptiness are numerous electric charges rushing about with great speed; but their combined bulk amounts to less than a billionth of the bulk of the table itself.'[1] No doubt this is technically correct, but the very strong implication that the 'scientific' table is somehow more 'real' than the 'familiar table', which is, accordingly, not really solid, is sheer nonsense, because it ignores the meaning of the word 'solid' as defined in common usage.

The fact that I cannot push my thumb through my table may or may not be due to, or conveniently explicable in terms of, the whirling electrons of my thumb bumping into or repelling the whirling electrons of the table; but this has nothing whatever to do with whether the table is 'really' solid or not. The table *is* solid ('really' being redundant, as usual) by definition; for the word 'solid' is *defined* (in effect) by saying 'such that you can't push your thumb through it'—and you can't. As for declaring, in another illustration, 'The plank has no solidity of substance. To step on it is like stepping on a swarm of flies';[2] this is not merely nonsense, it is flatly untrue—as anyone who has tried to step on a swarm of flies can testify.[3]

[1] *The Nature of the Physical World*, p. xii. [2] Ibid., page 342.

[3] For a much more competent criticism of Eddington's philosophical views than I can hope to offer, the reader should consult the late Dr. Susan Stebbing's book *Philosophy and the Physicists* (Methuen, London, 1937). The matter is important, because Eddington's great powers as a mathematical physicist (not yet perhaps fully appreciated) and his gift of picturesque exposition—which sometimes ran away with him, as in the above example—enabled him to exert over the lay mind an influence out of all proportion to his acumen as a philosophic thinker.

But Eddington gets into a much more serious muddle when he talks about Reality as a whole, or equivalently. He never seems able to get away from the conviction (as many passages show) that there exists something to be referred to as 'the external world', or 'the physical world'—which I suppose he would call the 'real' world—of which (he maintains) it is the aim of science to provide a symbolic description.[1] Knowledge of this world can only be inferential, and is obtained by a process of 'decoding' a 'cryptogram' formed of 'signals' transmitted by the nerves of the body to the brain, and there 'worked up into a vivid story' by the 'story-teller' who is 'the perceiving part of my mind'. This is, of course, the old 'appearance-reality' story over again, though more picturesque than usual, and disguised as science instead of metaphysics; and it is no more meaningful than any other version. But Eddington—more or less following Kant, if I am not mistaken—goes further, and insists that the mind not only decodes and interprets the signals, but is so constituted that it cannot avoid cooking the answer, so that '. . . what we comprehend about the universe is precisely that which we put into the universe to make it comprehensible'.[2]

I suspect myself that Eddington in these connexions is confusing the process of *abstraction*, whereby physicists restrict themselves to quantities which can be measured with a scale, a balance or a clock, and the alleged ensuring of the answers to our questions by the nature of what we '*put into* the universe'—which does not seem to me at all the same thing; but I may be quite wrong here. In any event, this business of first postulating (in effect) a wholly gratuitous distinction between Reality and Appearance, and then inserting

[1] Cf., for example, ibid., p. xiii. Apart from the improper personification of 'Science', this is simply not true. As Dr. Stebbing pertinently remarks, the aim of science 'can hardly be to express in language which only mathematicians can understand the occurrences with which everyone is familiar' (loc. cit., p. 66). The aim of scientists, I would rather say, is so to co-ordinate our experiences (cognitum sequences) as to enable us to predict with reasonable assurance what will happen under specified conditions, and thus, in particular, to enable us to perform whatever operations will lead to the results we desire. Or something very like this.

[2] *Relativity Theory of Protons and Electrons* (Cambridge University Press), p. 328. Quoted by Stebbing, loc. cit., p. 280.

a kind of automatic distorter in the form of the 'perceiving part of the mind' is a form of perverted ingenuity which cannot, I submit, stand up to intelligent criticism for a moment. If we do not and cannot know the truth, how do we know that the 'story-teller' is telling stories?

I'm sorry, but no matter how ingeniously we may wriggle, there is no getting behind the fact that all we can ever cognize are cognitum groups and sequences; and any inference we may make from these—whether we label it 'Reality' or not—must either be verifiable by the cognition of further groups and sequences, or else be discarded as inherently unobservable and therefore meaningless. Colloquial usage and irrelevant quibbles apart, Appearances are the only Reality.

54. *Reality and Existence, contd.* (4). The words 'existence' and 'exist' have given hardly less trouble than the words 'real' and 'reality'. The standard practice is to take it for granted that we know the meaning of the word 'exist', and to proceed at once to argue about whether bricks, glass mountains, centaurs, virtuous triangles, God, square circles, the State, the Will, or the square root of minus one, do or do not exist. Every so often someone takes time off for a fresh outburst of joyous wrangling about *Esse est percipi* (To Be is to be Perceived), and a good time is had by all—if they like that kind of thing.

Our old friend Parmenides took as 'the fundamental principle of his inquiry' the magnificent *dicta* 'Thou canst not know what is not—that is impossible—nor utter it; for it is the same thing that can be thought and that can be', and 'It needs must be that what can be thought and spoken of is; for it is possible for it to be, and it is not possible for what is nothing to be'. This sounded most impressive, and gave a great deal of trouble, until, after only a couple of thousand years, someone had the brilliant idea of saying that square circles, Centaurs, imaginary numbers, etc. (which obviously don't exist in the ordinary common-sense usage of the word), do not *ex*ist, but *sub*sist; and that, I understand, is the more or less accepted doctrine to-day.[1] Thus have the unremitting

[1] It cannot be said that subsistent entities enjoy any great popularity at present, among British philosophers at any rate. (H. H. P.)

labours of metaphysicians broadened and deepened the scope of human understanding.

Let us approach from a slightly different angle, by asking what we mean when we say 'X exists' or alternatively 'X does not exist'. We need not, I hope, go over all the ground again; the reader will doubtless recognize by this time that to assert that X exists is to assert that the referent of the symbol X is identifiable and allocatable to some 'order'—material imaginary, hallucinatory, oneiric, etc.; and that this referent needs defining by specification of what are ordinarily called its properties (cognitum sequences), if need be at the level of ostensive demonstration. The same is true of the word 'exist', and it seems clear to me that we cannot go deeper, so to say, if doubt is expressed, than to say, at this ostensive level, 'These cognita—coloured patches, tactile, auditory, thermal, etc., sensa—which you are now cognizing, *exist*'. I do not think anyone would be so intransigent as to maintain that whatever it may be that he is at any moment cognizing does not exist, unless he uses the term 'exist' as synonymous with (most probably) 'be material'; in that case, he might, knowing himself to be hallucinated, for example, declare that he was cognizing a 'non-existent' (i.e. non-material) Centaur. But then he is talking not about the cognita actually cognized, but about their relations to other cognita in a cognitum sequence.

But this is not particularly interesting. The question that has mainly perturbed philosophers in this connexion is not whether cognita (or 'ideas', 'impressions', 'sensations', 'percepts', 'sensa', etc.) exist, which nobody doubts, but whether it is correct to speak of *cognizables* existing when they are not, in fact, cognized. This has more usually taken the form of debating whether material objects do or do not exist when they are not 'perceived', but the difference is evidently only one of terminology. From the supposition that they do not has been drawn the conclusion that material objects (i.e. 'matter') are wholly 'mind-dependent', which leads to various forms of Idealism and Mentalism; while the supposition that they do is the basis of the more rational versions of Materialism.

Almost every philosophic work that purports to be at all comprehensive contains some considerable discussion of the problem; and it is one of the most curious phenomena of the subject that so much ink has been spilt on a point which is so easily settled. It is, as usual, only a matter of finding out what you are talking about, and what steps (if any) can be taken to ascertain whether what you are saying is true.

55. *Reality and Existence, contd.* (5). Suppose that I ring the bell for my butler and request him to bring me a whisky and soda—the fact that I have no bell, or whisky, or soda, and have never had a butler, being irrelevant. The well-trained minion leaves the room, closing the door behind him; but after a few moments he reappears, carrying the desired depressant. Obviously it is a matter of indifference to me whether he ceased to exist when I ceased to observe (i.e. cognize) him, and was magically recreated when observation was resumed, or persisted as a thought in the Mind of God, or subsisted as a society of Monads à la Leibnitz, or just went on being a butler in the crude everyday sense of the term—provided, of course, I get my whisky and soda. And, on the face of it, there is no means of distinguishing between the various hypotheses, since the end result is assumed the same in each case; and I think it would be pressing the conceit too far to suggest asking the butler.

Similar considerations apply, of course, to such simpler cases as whether my table still exists when I close my eyes or leave the room. Common sense, which is not always so bad a guide as philosophers have supposed, declares that it does; but it is not, at first sight altogether easy to see what evidence one can adduce in support of the declaration—for it would obviously be cheating to take a furtive peep to see whether the table is still there, as we all know it would be.

It is all a matter, I repeat, of deciding what we are talking about. The trouble has arisen from people neglecting to say how they propose to verify such statements as 'the table still exists, when unobserved' or 'is still there when we have left the room'. This is equivalent to defining the words 'exist' or 'is still there'; and there are only two courses open to us.

Either we must define the word 'exist' in such a way that we can ascertain which of the two contradictory statements, 'Under such and such conditions, X exists' and 'Under those conditions, X does not exist' is true; or else we must define the word in such a way that we cannot (or leave it undefined), in which case both statements become meaningless.

But first we must remind ourselves of what we are actually talking about. When we use the word 'table', we are, as we have seen, not referring to some mythological and inherently unobservable entity known as 'the table-itself', but to certain cognitum groups and sequences. No one, therefore, is entitled to dodge the issue by saying, 'Of course the table is there when I can't see it—I know that because I can *feel* it' (or *hear* the butler when he is out of the room, etc.). The question, normally phrased in terms of material objects, should properly be stated in terms of cognita, i.e. Do cognita exist (as 'cognizables') when not cognized? or, Are there uncognized cognizables? No one doubts that the tactile table exists when the light is out—one can touch it, bump into it, and so forth—but what about the visual table? I sit at my table in the ordinary way, and open and close my eyes, or turn the light on and off, several times. I alternately see and do not see the table, i.e. cognize and do not cognize the relevant visual cognita, though the tactile cognita corresponding to the statement, say, that my elbow remains in contact with the table, are substantially unchanged. What happens to the visual cognita when my eyes are closed or the light is out?

There are two and, I think, only two, alternative hypotheses which might be advanced: one is that visual cognita continue to exist as cognizables, the other that they cease to exist but are recreated so soon as I open my eyes or the light comes on again. But the observable facts would, *ex hypothesi*, be exactly the same in each case; the so-called alternatives are, therefore, inherently indistinguishable; the alleged distinction is, therefore, meaningless; the two verbally different statements are, in fact, saying the same thing; the phrases 'continue to exist' and 'cease to exist but are recreated' are identically equivalent; they do not refer to two different events, processes, or states of affairs, and it is meaningless to say that they do.

Strictly speaking, there is nothing to prevent us adopting the extinction and recreation view, if we insist on doing so; for it involves no contradiction in terms, nor, so far as I can see, any incompatibility with observable fact. But, as in the case of the *Ding*, we can do so only on condition that we make no use of it; and we should at once be faced with unanswerable conundrums about how and why and by what agency the cognita were extinguished and recreated. We should be much in the position of the White Knight's aged gate-sitter, who, it will be remembered, '. . . was thinking of a plan To dye one's whiskers green, And always use so large a fan That they could not be seen';[1] or of a man who is convinced that he has a thousand-pound balance at the bank, with the trifling reservation that he inexplicably cannot use it either to cash cheques or support credit.

With all respect, and the whole history of metaphysics notwithstanding, no enlightenment can proceed from producing gratuitous difficulties with one hand and unintelligible answers to them with the other.

We, therefore, say firmly that the entities known as cognita when they are cognized exist also as cognizables when they are not; and we define the word 'exist' in such terms as to make it so. This is just as compatible with observable fact as the only alternative, it avoids all complications about the mechanisms of extinction and recreation, and is in line with our common-sense treatment of situations involving material objects. If I attempt to raise money on the security of a lump of gold or a sheaf of securities, allegedly kept in my safe but unfortunately never to be found when the safe is opened, the stony-hearted money-lender decides that the situation is empirically indistinguishable from there being no or gold securities, and only his glass eye weeps when I tell him a pathetic story about the wicked fairies responsible for magic-ing them away whenever he calls to see them. This

[1] Lewis Carroll, *Through the Looking-Glass*, Ch. VIII.

corresponds to the case of the *Ding* (or any other unobservable) which we accordingly say does not exist. And conversely 'only philosophers with a long training in absurdity', as Russell has it, think it worth while to discuss whether an object which can always be observed in its proper place, given proper conditions, continues to exist between occasions of observation.[1]

To exist, then, is not 'to be perceived' (i.e. cognized), but to be cognizable; and to be cognizable is to exist.

56. *Reality and Existence, contd.* (6). The importance of clearing up this point, which is solely a matter of how we propose to use words, lies in the fact that it enables us to satisfy the central demand of all reasonable versions of Materialism, which has certain political advantages, as one might call them, if nothing else.

It seems to me legitimate to distinguish somewhat sharply between the crude and noisy materialism characteristic of some of the Victorian physicists and of the earlier years of the present century, and the more rational versions which appear to be gaining ground to-day. To the former category I relegate as unworthy of serious attention all who affirm that 'Matter is the only Reality', or deny 'the Reality of Mind' or maintain that 'Consciousness is an Illusion', or like folly. Such remarks indicate no more than a dogmatic refusal to face the facts of what goes on when they perceive, imagine, or (as they would doubtless claim) 'think', coupled, of course, with a stubborn ignorance of the functions and mechanism of language. Their almost classical slogan, 'The brain secretes consciousness as the liver secretes bile', contains perhaps more absurdity per word than any would-be serious remark ever made, with the possible exceptions of 'The divine is rightly so called' and Tertullian's *Credo quia impossible.*

But sanity seems to be gaining ground, and the gulf between the materialistic view and the correct one, like that between this and the idealistic, now seems not too wide to bridge. For

[1] Cf. Ayer, *The Foundations of Empirical Knowledge*, p. 67, who writes: 'The criterion by which we determine that a material thing exists is the truth of various hypothetical prepositions asserting that if certain conditions were fulfilled we should perceive it.'

example, my friend Professor J. B. S. Haldane—whom I should describe as not merely an eminent materialist, but an eminently sane one—has suggested to me that the essential characteristic of materialistic doctrine is 'belief in something not dependent for its existence on our knowledge of it'.

So far as the existence of something not dependent for that existence on our knowledge of it is concerned, cognizables obviously fill the bill to perfection. For, as we have just seen, it is not only legitimate but almost compulsory to regard them as existing when they are not cognized (known); so the central tenet of the faith is comfortably covered. Moreover, as I have already indicated, I am all in favour of extending the scope of physics—by which I mean the observational and mathematical methods employed by physicists—to cover the whole range of phenomena of all orders, whether material, mental, spiritual, or otherwise; and whether you then choose to bracket them all together as 'physical', or not, is merely a question of convenience. And there can be little doubt that such an extension of physics will involve the inclusion in it of processes, relationships, etc., immensely more complicated than any it yet considers.

There is accordingly no difficulty in conceding the main point demanded by the saner varieties of materialist. Where they go astray is in supposing that the 'something' which exists independently of our knowledge of it is *matter* (or 'consists of' matter). It is not. It is or consists of *cognizables*, which exist as such independently of whether they are or are not related to other cognizables (see the discussion of Mind, below, Ch. V) in that way which constitutes being cognized.

Anticipating the course of the discussion to some extent, it cannot be too strongly insisted that cognizables or cognita are no more intrinsically material than they are mental, or spiritual or anything else; and it is meaningless to say that they are 'made of' anything, since any attempt to answer the question 'Of what are cognizables made?' would have to be couched in terms of observations of some kind, themselves consisting of cognitum groups and sequences.

It is interesting to note that the classical Idealists, notably

Berkeley and his followers, have fallen into exactly the same error (or perhaps, rather, type of error) in their approach from the opposite direction, namely, that of supposing that the entities of which we are directly aware must be *mental*. They are not; they are just cognita. Thus Ayer writes that Berkeley '. . . maintained that to say of various "ideas of sensation" [cognita, notably sensa, W. W. C.] that they belonged to a single material thing was not . . . to say that they were related to a single unobservable underlying "somewhat", but rather that they stood in certain relations to one another. And in this he was right. Admittedly he made the mistake of supposing that what was immediately given in sensation was necessarily mental; and the use by him and by Locke of the word "idea" to denote an element in that which is sensibly given is objectionable, because it suggests this false view. Accordingly we replace the word "idea" in this usage by the neutral word "sense-content", which we shall use to refer to the immediate data not merely of "outer" but also of "introspective" sensation, and say that what Berkeley discovered was that material things must be defined in terms of sense-contents.'[1]

The point is that the two parties have fallen into similar and almost symmetrical errors of supposing on the one hand that 'that which exists independently of our knowledge of it', and on the other that 'that which is the immediate object of consciousness' (or like phrase), is necessarily whatever it is in which they happen to be chiefly interested, namely 'matter' (or material objects) and 'mind' (or mental objects) respectively. Neither party is right; both are wrong. The independent existents, and the immediate content or objects of

[1] *Language, Truth and Logic*, pp. 53–4. Although I wholeheartedly agree with Ayer's contentions, I do not like his terminology very much better than that of Berkeley. The term 'sense-content' seems to be to carry almost as strong a suggestion of 'sensation', involving receptors, nerves, brains, and other furniture of materialism as Berkeley's 'ideas' carries suggestions of mentalism (cf. Sect. 42, above, p. 101); and although he qualifies it by making it include 'introspective' sensation, it is not at all clear whether this is intended to cover, as it should, the constituents of imagery, dream, hallucination, etc., or to refer only to sensations proper of endosomatic origin.

consciousness, awareness, etc., are alike those irreducible entities which I call cognita when they are cognized and cognizables when they are not. As Russell says, 'Physics and psychology are not distinguished by their material. Mind and matter alike are logical constructions; the particulars' [cognita 'out of which they are constructed, or from which they ar inferred, have various relations, some of which are studied by physics, others by psychology.'[1]

This enables us to clear up once and for all—at least for all who have ears to hear—the long-standing Materialist-Idealist controversy, which is seen to be futile so soon as it is realized that the words 'real', 'reality', etc., are always redundant, that the word 'exist' must be defined in empirical terms, and that the two schools differ only in talking about different con figurations of the same entities. Thus the Idealist is saved from the embarrassment of having to admit that he has barked his shin on an unreal bucket; while even a Dialectician given time, may be brought to realize that he cannot ulti mately refer meaningfully to anything but cognitum group and sequences.[2]

57. *Note on Space and Time.* I cannot do more than pause for a moment here to close one possible loophole of which critics might otherwise avail themselves, namely, of saying that although my reduction of all material objects, and there fore of matter generally, to terms of cognitum sequences is undeniable, I have ignored the nature of Space and Time *in* which, as they would contend, material objects and event respectively exist and occur.

For the purposes of our present discussion I think th

[1] *Analysis of Mind*, p. 307. As I have remarked above (cf. Sect. 50) am never quite happy about saying that anything *is* a 'logical construc tion'; I would say rather that material and mental objects alike *ar* cognitum sequences of differing types of pattern, or words to this effect and I do not much like the idea of 'inferring' matter (say) *from* particular (cognita); inference comes in, I think, only in the sense that cognition o certain types of cognitum sequence leads us to expect, by virtue of pas regularities, certain other types of cognitum sequence. But the main point is brought out clearly enough, namely that the difference between matter and mind is not one of differing 'substance', or any equivalent but of the difference between the 'various relations'.

[2] Cf. Ch. VI, Sect. 88, pp. 230–1. (H. H. P.)

answer is fairly simple, at any rate so far as space is concerned. It is to the effect that to speak of material objects being 'in' space, in any sense which implies or takes for granted that space could exist without material objects, or, more obviously, material objects without space, is to indulge one of those artificial dichotomies which are often linguistically convenient but in fact devoid of meaning. It is easy enough to imagine, or rather to fancy that we are imagining, something that we would call 'empty space'; but, if we rigorously exclude from our image all trace of colour, shading, etc., and of our own presence, we find that we are (quite correctly) imagining exactly nothing at all. Truly empty space would be inherently and *ex hypothesi* unobservable. It would be literally impossible to verify any statement whatever about it, including the statement that it exists; and it must accordingly be ruled non-existent.

We cannot meaningfully divorce space from matter. Any verifiable statement about space must involve some mention of matter, or material objects, express or implied; and these statements must be reduced to terms of cognita. Even if it be claimed that this is true only of 'physical' space, and that it may be meaningful to speak of 'non-physical space', which I doubt,[1] the inevitability of ultimate reduction to cognita would still hold. Admittedly, many cognita (coloured patches, etc.) have 'extension', which is commonly reckoned a spatial quality; and Price and others speak of 'visual depth' as directly given in experience.[2] But the fact that a cognitum may properly be described as 'extended' or 'deep' is no justification for saying that it exists 'in' a space which could in any sense itself exist without it. Delete all cognita and you cognize nothing. Some cognita are extended, some are deep, some may be both for aught I know, just as some are yellow, some acid, some shrill, etc.; but all are cognita.

[1] What of the space of dreams or hallucinations—or even of visual images? (H. H. P.)

[2] *Perception*, p. 3, already quoted, and many other passages. Personally, I do not seem to be able to identify this alleged quality of 'depth', but that is presumably my fault. Certainly there is a good deal of modern work which seems to show that the function of binocular vision is not, as was once supposed, to give us the experience of solidity but the experience of flatness.

Personally, I think it would be best to restrict the term 'space' to what you can measure with a scale (or, more sophisticatedly, to some system of relationships constructed out of scale measurements, or the like), i.e. to physical space. If we find that cognita organized in non-material patterns exhibit relationships (e.g. of relative accessibility or remoteness) analogous to spatial relationships, then it would probably be wise to adopt some new term by which to refer to these, thus avoiding possible sources of confusion.

But in either case, whatever statements we may make about space must either be verifiable or meaningless; and if they are verifiable, then the verification must involve the making of observations of some sort, and these must necessarily consist in the cognition of cognitum sequences of some kind—notably, in the case of physical space, those of eye-movements and accommodation, of stretching, walking, touching, noting coincidences of marks on scales, etc. So my account does not, as the critic might suppose, fail to cater for the space *in* which material objects exist or events take place.

Time is somewhat trickier, but the same principles apply. It is just as meaningless to talk, directly or by implication, of 'eventless time' as it is to talk of 'empty space'—and for the same reason, namely, that it would be inherently unobservable Yet I have heard even the most eminent philosophers gravely discussing, to my horror and amaze, whether Kant[1] (I think) was right in maintaining that, under certain conditions or on some hypothesis, something-or-other must be regarded as 'filling time', on the ground that 'time must be filled with something'—or words to a similar effect—for all the world, apart from the actual wording, as if Time were a sort of endless tube in which Nature would abhor to see a vacuum! Change and time are, I submit, synonymous or virtually so. Freeze everything, so to say, including the flux of your own mental states, and time stops. If there is *no* change there is *no* time.

I must confess that, at the moment, I regard—not quite contentedly—succession and change (the terms being again

[1] Cf. *Critique of Pure Reason,* The Analogies of Experience. (H. H. P.)

virtually synonymous) as an irreducible brute fact which I see no possibility of 'explaining'; but that it falls into the category of cognita (or of relations between them), and that it is non-sensical to speak of Time as an entity *sui generis* capable of 'existing' independently of them is, I think, quite beyond dispute.

58. *Review and Summary*. We may summarize the foregoing discussion of Matter approximately as follows:

We clear the ground by saying that the word 'matter' is a generic or class name for all material objects[1] and concentrate on examining what is entailed by such a statement as 'This is such-and-such an example of a material object'—e.g. 'This is a lemon.'

To do this we study the typical perceptual situation, which we find to be essentially a Sign-situation. Certain immediate objects of awareness (yellow patch, etc.), commonly called 'sensa' but subsumed in my terminology under the more general term 'cognita', are interpreted, as a result of the contexts of which they have previously formed parts and of their present context, as *signs* that certain other cognita (e.g. tactile) will be cognized under certain conditions; in particular that these will follow certain yet other 'operational' cognita, such as the kinaesthetic cognita of stretching and grasping, etc. If these expectations are realized we say that, so far at least, our interpretation was correct, and that the object conforms to the defining specification of a natural and material lemon. But if not, then it is not a lemon, but an hallucination, or perhaps a material object of some other kind (e.g. a wax or plaster imitation), according to circumstances, and we have misinterpreted the signs.

No other procedure will enable us to verify or refute any statement that may be made about the object, and no test can

[1] Perhaps I should have brought out more clearly than I did that, if anyone objects to dismissing the word 'matter' in this way, and wishes to use it as synonymous with 'substance', then everything that has been said about the inherent unobservability of the *Ding*, the impossibility of verifying any proposition about it, and its consequent non-existence, applies with exactly equal force, *mutatis mutandis*, to 'substance'—which might, indeed, be defined as 'that of which "things-themselves"' are composed.

possibly yield or consist of anything but alternating sequences of 'operational' (kinaesthetic, etc.) and 'observational' (visual, tactile, etc.) cognita or groups thereof. It is accordingly meaningless to maintain that a material object consists of anything other than cognitum sequences, for anything else is of necessity inherently unobservable, so that no statement, even to the bare effect that it exists, can be verified.

In particular, it is meaningless to speak of 'things-themselves', *Dinge an sich*, or to use equivalent words purporting to refer to unobservable entities alleged to 'have' properties, and all language of this kind is to be ruthlessly barred, except on the strict understanding that it is used as a kind of shorthand purely for the sake of linguistic convenience.

The chief current 'theories', so called, of Perception are briefly indicated, and it is noted that there is no possibility of adducing empirical evidence in favour of one rather than another, or of deducing verifiable consequences from any of them; at best, therefore, they may be regarded as alternative languages, though it seems doubtful whether rules could be given for translating from one into another. It is suggested that a more positivist statement—e.g. that in appropriate circumstances the visual stick *is* bent, the visual penny *is* round, etc.—might be preferable, and might be held to rank as a genuine theory, in so far as it would enable us to deduce the possibility of hallucinations, etc., even if we have never experienced them.

The words 'real', 'really', 'reality', are found to be invariably redundant, and therefore liable to mislead, since 'real' can never mean more than 'conforming to specification', i.e. is always replaceable by this phrase. All cognizable objects (using the term in the widest sense) are 'real' in their own order—material, dream, hallucinatory, imaginary, etc.

The word 'exist', also, must be defined in observational terms, like any other. The criterion of existence is that the entity in question should be observable (cognizable) under appropriate conditions. No process of observation can (by definition) verify the alleged existence or non-existence of an uncognized cognizable. To maintain that a cognitum ceases to exist when not cognized would entail the gratuitous postulation of an *ad hoc* mechanism for its extinction and recreation. It is accordingly legitimate, and virtually obligatory, to regard cognita as existing (as 'cognizables') when not in that relation with other cognita which constitutes their being cognized.

This enables us to meet the basic requirement of sane materialism without conceding non-sensical contentions to the effect that 'Matter is the only Reality', etc.

All materialists err in supposing that whatever it is that exists is necessarily 'matter'; all idealists err in supposing that the immediate content of consciousness, or objects of awareness, are necessarily 'mental'. In each case the entities in question are cognizables, uncognized or cognized, and these are not intrinsically either material or mental, but neutral.

All statements about space and time can be reduced to statements about cognitum sequences, in the same way as statements about material objects.

V
MIND

59. *General.* In certain respects we shall find it easier to deal with 'Mind' than with 'Matter', for most of the heavy logical work has already been done; but there is one respect at least in which this is not the case. We start, of course, by saying that 'Mind' is a symbol used to refer to mental objects in general (or possibly, but erroneously, to the alleged underlying 'substance' of which they are composed), in the same way that 'Matter' is used to refer to material objects, etc.

There is, however, an important difference between the two cases, inasmuch as we cannot fall back, in the second, on ostensive demonstration to clear up points of ambiguity. If someone declares that he does not know what we mean by the term 'material object', we can coercively demonstrate with the aid of tables, chairs, books, and bricks—or sticks, boots, and fists if he prove obdurate; or if he says he cannot understand the word 'yellow' we can show him a series of yellow objects and consign him to an asylum if he proves incapable of making the simple abstraction required. But if he demands that we show him a mental object—an image, a dream, an hallucination—we are not so fortunately situated. In special cases, I suppose, we might do something with the aid of hypnosis, but it is not in general practicable to define mental objects and events ostensively in the way we can apply to material objects.

Again there is always the possibility that some hardened sinner may seek to maintain that he is the only conscious being, and unique in enjoying mental states, etc.; or, worse, affirm that he is not conscious at all and does not know what we mean by saying that we are so. Such contentions are by no means easy to overset, and Ayer makes an important point when he argues (*Language, Truth and Logic*, pp. 200–2) that it is not legitimate to use 'an argument from analogy, based on the fact that there is a perceptible resemblance between the behaviour of other bodies and that of my own, to justify a belief in the existence of other people[1] whose experiences I could not conceivably observe'. His objection is the perfectly sound one that if other people's experiences are inherently unobservable by me, then, so far as I am concerned, they are 'metaphysical objects', all assertions about which are bound to be meaningless, since none can be verified or disproved. I am not altogether happy about Ayer's method of getting out of the difficulty by defining other people's experiences in terms of one's own, and personal identity in terms of bodily identity, thus evading the assumption that other people's experiences are not experienceable by oneself; and it seems to me that this assumption can almost certainly be negated by appeal to the facts of telepathy, which must now be accepted as genuine phenomena. I shall refer to this again later.[2]

But whatever may be the best way of coping with this sort of conundrum, I think it would be only waste of time to embark on a long discussion of it here. I do not suppose that objections on these lines will be raised by any serious critic, and I think I shall be safe in taking it for granted as agreed that other people do have minds more or less closely resembling that to which I refer when I speak of my own mind; that they are conscious in much the same kind of way that I am; that they dream dreams, experience images, are occasionally liable to hallucination—and so forth. Naturally, there will be wide divergences; some people hardly dream at all, and some only in monochrome when they do, while others dream copiously and in full colour at that; most people, so far as one knows, have never experienced an hallucination at all—I don't think I have—but some unfortunates seem to experience little else; in some, visual imagery predominates, in others auditory or even kinaesthetic, while some seem to enjoy hardly any, so

[1] Since he clearly is not talking about other people's *bodies* here, I think he must be referring to what would ordinarily be referred to—putatively at least—as other people's minds. [For another, and less radical, view concerning 'other people's minds', see Professor Ayer's later work, *The Foundations of Empirical Knowledge*, pp. 168–70. (H. H. P.)]

[2] See Sects. 78–9, below, pp. 201–7. (H.H.P.)

far as can be ascertained from what they say, though a few again can command images so vivid as to be barely distinguishable from actual scenes.

But such differences as these, or any others that might be mentioned, are beside the point; and I propose to by-pass any objections that might be raised here, and to go on to discuss what, if anything, we can properly be held to refer to when we use such terms as 'mind', 'my mind', 'consciousness', 'self-consciousness', 'self', 'ego', 'soul', 'spirit', and the rest.

60. *Non-existence of the Ego.* Let us begin with the last group in the list, where the crux of the matter lies.

The first and all-important point to be grasped in considering 'the mind' is that everything whatever that I have said about the 'thing-itself' or *Ding* applies with equal force to the 'mind-itself' or *ego*—or the 'I' or 'soul' or 'self' or whatever else you like to call it. In a nutshell, you cannot cognize anything but cognita; any alleged entity other than cognita is, therefore, inherently unobservable; all propositions concerning any such entity are, therefore, inherently unverifiable; it incessantly and necessarily eludes all attempts to confirm its alleged existence; it necessarily fails to satisfy the criteria by which alone we can decide that anything at all exists; it must, therefore, be ruled non-existent.

This is another and flagrant case of the domination of linguistic forms over our thinking. We habitually use forms which imply that there exists some entity (the *Ding*) which *has* the properties which we ascribe 'to' material objects, and unthinkingly conclude that there must therefore be such an entity, whereas there are only the properties (cognitum groups, etc.). Similarly, we habitually use forms which imply that there is an entity (the 'I', etc.) which knows, perceives, cognizes, etc., and similarly, conclude that there must be such an entity. But it is as elusive as the rainbow's end, and no amount of introspection enables us to catch up with it. Nor could it; for all that the most sedulous introspection could ever yield would be fresh groups of cognita.[1]

[1] [Cf. Hume: 'For my part, when I enter most intimately into that I call *myself*, I always stumble on some particular perception or other,

It follows that anything to which we may find it convenient to refer by such terms as 'mind', 'self', 'ego', 'I', 'soul', 'spirit', etc., must be built up, so to say, of cognita and cognitum sequences, just as anything which we find it convenient to describe as 'a lemon', 'a brick', 'a stove', etc., must be. I need hardly say, I hope, that this is not intended to imply that the patterns in which the cognita forming minds, mental objects, etc., are built up, or arranged, or the regularities which characterize them, are identically similar to those we find in the case of material objects. On the contrary, the whole point is that, whereas the constituent elements of material and mental objects, phenomena, or events are the same, in the sense of being alike cognita,[1] their arrangements, patterning, sequences, and regularities are different—in the one case those we codify under the name of the Laws of (material) Physics, in the other those we codify under the name of the Laws of Psychology. If the visual lemon-cognita, so to call them, when followed by the appropriate kinaesthetic cognita (stretching, grasping), lead on to the tactile cognita appropriate to the defining specification of a material lemon, we conclude (provisionally at least) that we are dealing with a material lemon; if they do not, then the object is not material but hallucinatory, imaginary, oneiric, or the like, i.e. some sort of a 'mental' lemon.

As a start, then, and broadly speaking, we separate off, so to say, from the total mass of cognizables those which, conforming to the laws (exhibiting the regularities) of material physics, constitute the material, or, as we would often say, the 'external' world; all others we lump together, provisionally, and speak of as forming the mental world or world of mind. The term 'mind', used in a wide inclusive sense, refers to all those cognita and cognizables which do not exhibit the regularities characteristic of material objects. This is very rough

of heat or cold, light or shade, love or hatred, pain or pleasure' (*Treatise of Human Nature*, Bk. I, Pt. IV, Sect. 6; Everyman Edition, Vol. I, p. 239). (H. H. P.)]

[1] In case I have not yet made it sufficiently clear, I should like to emphasize here that the use of the word 'cognita' is to be understood as including 'cognizables' wherever the context requires it.

and ready, for it ignores the fact that, as we shall see, there is a considerable overlap between the two classes, and it evades the question (which is, as a matter of fact, purely linguistic) of to whom or to what the regularities are exhibited.

Two points may be noted here. First, although I have provisionally used the single word 'mental' to refer to the whole class of non-material cognita and cognizables, there is no reason in principle why this class should not be subdivided into as many sub-classes as study of the facts may in due course warrant. It may be that when we have acquired as much knowledge of the laws of psychology as we now have of the laws of physics (or preferably more) we shall find certain types of regularity (laws) separating themselves out, so to say, from the welter of cognized sequences in much the same way that the regularities of the material world have done; and we may then find it convenient to confine the term 'mental' to these, and to adopt another term, such as 'spiritual', for use in connexion with the others.

Second, I am inclined to think that the division between 'material' and 'mental' is to a great extent arbitrary, and more the product of ignorance than of knowledge. That is to say: if we understood the mechanism of hallucination, say, as well as we understand the mechanism of certain material phenomena, so that we could predict or induce hallucinations as certainly as we can predict or induce explosions or electroplating, I suspect that these would be reckoned as much as they in the category of physical events; for this, very approximately is what we mean by a physical event. And it is interesting to note—though not, I think, more—that certain Indian philosophers have grouped some of the 'lower' mental processes along with the material as ordinarily understood, in a single category of 'matter' known, I believe, as 'prakriti'.[1]

61. *Rough Sketch of a Mind.* No one, unless he explicitly claims to be unconscious and to enjoy no experiences whatsoever—in which case he can, I think, be dismissed as merely

[1] The reference is probably to the *Sankhya* philosophy. See Sir S. Radhakrishan, *Indian Philosophy* (London, Allen & Unwin, second edition, 1930), Vol. II, Pt. III, Ch. IV. (H. H. P.)

a perverse nuisance—can deny the existence of cognita, or (I submit) of cognizables. And I doubt whether anyone will quarrel with the statement that some of these behave, so to say, in that relatively orderly manner characteristic of matter (material objects and events), while others do not. It is accordingly easy enough to accept the broad notion that 'mind' consists of all cognita and cognizables not forming part of material objects.[1] But it is not quite so easy when we come to inquire about what we are referring to when we speak of 'my mind', or 'your mind' or 'Smith's mind'. What distinguishes yours from mine, or either from Smith's? Where does one leave off and the other begin? According to what principles, or 'forces', etc., are cognita, not co-ordinated as parts of material objects, organized into what we call 'individual' minds? What is it that earmarks, so to say, a given cognitum as being a part of your mind and not of mine? Or are all or any of these questions senseless?

I can, of course, only speak at first hand of my own mind; but, as I have said, it seems reasonable—for practical purposes at least—to suppose that other people have (as the phrase is) minds not wholly dissimilar from my own, and I shall allow myself this latitude, even though we cannot lift the supposition into the realm of verifiability till a somewhat later stage. And, of course, I am obliged to rely, especially in the early parts of the discussion, on more or less traditional forms of words, such as, taken at their face value, tend often to beg the questions at issue. Let us none the less attempt some examination of that queer phenomenon I call my mind at any given moment, such as the present.

In ordinary colloquial language, I sit at my table and intermittently hammer the typewriter, pausing every so often to think for a moment, to relight my pipe, or to appease the mewings of the cat. If we expand this into terms of what is actually going on, we should have some such account as this: There is what we commonly call a 'field of consciousness'—the term 'field', it is to be understood, being used rather in

[1] Subject to the reservation about 'overlap' already mentioned and to be discussed later.

the same sense as in 'a field of force' than in that of a sharply delimited piece of land. Within this field, cognition is occurring. This is not an ideal way of putting it; but the language does not happen to have developed in a way ideal for this sort of purpose. Ordinarily we should say that '*I* am conscious *of* this and that', or 'I am cognizing so-and-so', but I want to keep the 'I' out of it as much as possible. Roughly speaking, the field—i.e. the cognita concerned in the situation—consists of certain visual cognita, notably the black and white patches of the typewriter keys, the paper and the typed letters as they appear, with many others (table, blotter, calculating machine, card indexes, hands, clothes, etc., etc.) less central, rather marginal, less, as we say, attended to; there are also tactile cognita as my fingers tap the keys, and from the pressure of my body on the chair or my feet on the ground. Very important are successions of verbal images, part auditory, part kinaesthetic, as the words about to be written are juggled and rejuggled, the sentences cast and recast—or, occasionally, flow without preliminary effort. Though there are certain fairly constant components, the field is never the same for two successive moments; it is in a state of perpetual flux and change, with sometimes one set of cognita predominating, sometimes another; and sometimes some quite irrelevant group (e.g. mews from cat) breaking in, as it were, from outside, or others (sensations of thirst, or nicotine craving) arising from within the body. Maybe, again, when inspiration (if that's the word for it) falters, some intrusive group reminds me, as we say, of some extraneous matter, such as an overdue letter, and a whole host of memory images crowds into the field, displacing all too easily the sequence of verbal images which I should roughly describe as 'what I am thinking about and want to write'. Often enough, wearied with my efforts and dissatisfied with their results, I come to a stop, and the field is filled with cognita describable as forebodings of failure ever to get this book finished or to make people understand what I am trying to explain, accompanied by those unpleasant sensations mainly of visceral origin (sinking feelings, etc.) with which all who are liable to depression are only too familiar. However, this is amply sufficient to indicate what I am talking about when I speak of a 'field of consciousness' or a 'state of mind', and I have no doubt that, altering the details to suit each case, everyone will recognize the description.

62. *Sketch of the Mind, contd.* It is obvious, however, that there is more to a mind, so to say, than the content of a field of consciousness at any moment, as the phenomena of memory alone are sufficient to show; and the reader may well be wondering also how I am going to deal (since I deny the commonly postulated 'ego' as a metaphysical nonentity) about *what* it is that is conscious *of* the field or its content. With some hesitation, I think it will be preferable to take the first point first.

I have never been able quite to make out what the ordinary or orthodox view of the mind is—or indeed whether there is one at all. So far as my impressions go, it is most usually thought of as a kind of container or receptacle, itself unobservable, in which are in some sense located the images, etc., of which we are from time to time said to be conscious; and somewhere about the place there is a kind of bull's-eye lantern, occasionally operated by the Will (a sort of psychic policeman), which is turned hither and thither to illuminate this bunch of images or that. Or, alternatively, it may be discussed, as with such immense ability by Professor Broad, in terms of Substance and Attribute,[1] with all the disadvantages attendant upon using the word 'substance'—substances being inherently unobservable. Or, again, we have the psycho-analytic scheme, with all the emphasis laid on a kind of stratification—the conscious, the sub-conscious, the fore-, pre-, and un-conscious, etc., with censors at every level to prevent the free circulation of ideas (just like real life, one might almost say).

The above remarks are, of course, mostly libellous and inserted mainly for the purpose of suggesting, as I think is true enough, that existing notions about the nature and mechanism of the mind are vague where they are not

[1] Cf. *The Mind and Its Place in Nature*, particularly pp. 588 ff. and 610 ff.

over-concretized, divergent, and confused. Apart from the analysts, broadly speaking, only the metaphysicians (whose remarks are meaningless from the start) have had the courage to talk about the mind at all, for professional psychologists, whose business it is to consider it, have been far too prone to tether themselves to the physiologists' apron strings in the hope of acquiring scientific respectability.

But I want to pick out two points, which seem to me to be instructive, from the muddle of contemporary notions. First of all, *why* do people cling to the 'container' conception, and continue to talk of this and that being 'in' the mind? Largely, of course, it is a matter of verbal habit; but I fancy there is more to it than that. I think it probable that there is another, purely psychological factor in the form of a feeling that, unless the content of the mind is in some way limited, it will somehow cease to be one's own private and particular preserve, and thus our cherished sense of individuality will be imperilled. Probably the same kind of fear is at the root of the second feature, namely, the notion of consciousness or attention being focused or concentrated 'on' one portion or another of the mental content. The thoughts, feelings, sentiments, images, etc., which fill the field of consciousness from moment to moment are, as we all know, fleeting and evanescent in high degree; and I think we[1] feel a kind of self-preservative urge to postulate something more permanent than they, something that shall be, as it were, their master—a Thinker of the thoughts, an Imaginer of the images, a Cognizer of cognita, above all, a Something that shall still be ourself even all these pass away into nothingness.

None the less, and whether we like it or not, every such notion of any kind of a Mind-itself, however verbally disguised and whether conceived as a container, a limiting perimeter, a thinker, an illuminator, a substance, or a Pure Ego, in so far as it is supposed to be something other than cognita, is inherently unobservable, essentially metaphysical, and wholly devoid of meaning. We can only meaningfully

[1] [What does the word 'we' mean here? Should the author have said 'there is' instead of 'we feel'? (H. H. P.)]

discuss mind in general, or 'individual' minds in particular, in terms of cognita and cognitum sequences, for exactly the same reasons as make this true in the case of matter and material objects.

Since this is the case (and it cannot be emphasized too strongly) I should only be wasting the reader's time and my own if I were to discuss the various difficulties to which such views give rise, or the solutions which have been proposed for dealing with them. They are all pseudo-difficulties and false solutions, inasmuch as they all arise from postulating some sort of an Ego or Mind-itself conceived as capable of existing independently of the cognita which would ordinarily be said to form its content. It will be more profitable to see what kind of a picture we can form of the mind using nothing but cognita, their groupings, and their sequences. It all seems to me to be, in outline at least, so simple and straightforward—once we have got rid of the metaphysical monstrosities, and are prepared to forgo some of our unanalysed wish-thinkings—that the chief difficulty is that of persuading oneself and others that nothing more elaborate is necessary.

63. *Sketch of the Mind, contd.* (3). In the absence of a highly developed mathematical technique, which we have not at our disposal—and which neither I nor the reader would understand if we had—it is impossible to discuss a matter of this kind without indulging in a certain amount of model-making and what I might call 'hylomorphism', i.e. thinking of something that is not material in terms of things that are; and this is bound to be in some degree misleading. Still, provided we do not allow ourselves to argue from illegitimately imported properties of the model or the material objects, we need not come to much harm.

Let us for a start conceive of 'mind' generally as an immense assemblage of discrete particles—rather like a nebula—the particles being of different kinds, and perpetually in a state of motion under the influence of 'forces' the nature of which it is our business in due course to ascertain. The particles are, of course, cognita, and their differences must be supposed such as to cover every way in which one cognitum may differ from another, e.g. dark red from light red or either of these from any shade of blue, or any colour patch from any noise or pressure or taste, etc. Even at this point we must pause to reflect that, whereas it is impossible to imagine such an assemblage except in space, this at once introduces a dangerous distortion. Cognita do not exist in space; space is made of cognita, so to say, in the sense that any statement about space must either be reducible to statements about cognitum sequences, etc., or else be meaningless.[1] This well illustrates the point I made a few lines ago; it would be quite incorrect to argue (as might easily happen) that because, in our imagined assemblage, one particle was spatially more remote from another than a third, there would therefore be any difference in the ease with which these might interact or influence each other.[2] Similarly, it would be dangerous to take too literally the discreteness of the 'particles'. Speaking personally, my own image is a kind of cross between a nebula, a cloud of midges, a perpetually shaken kaleidoscope, and the shifting colours on a soap film, but that is of no importance to anyone else.

Now imagine that there are formed certain condensations, or clusterings, of the particles in the mass, but let them not be either too compact or too homogeneous; let it be rather a matter of more or less gradually and, particularly, continuously increasing and decreasing density round certain points, somewhat as the density of population increases and decreases as one approaches the centre of a city from the open country, passes through it, and goes out into the country again. Indeed, one may conveniently change the whole image in this connexion and substitute one of people and cities, if that is found easier to think of. These condensations represent what

[1] It is true that certain cognita, notably visual, are 'given' as extended, while others are not; but it would be rash, to say the least of it, to identify this extension with the physical space in which we find ourselves obliged to visualize our congeries of particles. Cf. Ch. IV, Sect. 57, above.

[2] Mathematically, I suppose, any difficulty of this kind would be overcome by assigning a different dimension to every particle—there is nothing against doing so—or at least to every distinguishable type.

we call individual minds, and the region of maximum density represents what we call the field of consciousness at any moment—which, as we know, is not sharply delimited but shades off, as it were, from those cognita of which (as we say) we are most vividly conscious, through others of which we are less so, to some which we call marginal and of which we are barely conscious at all.

Let us pursue this image a little further, again remembering that we must not allow ourselves to be led astray by irrelevant features, as that men live in brick boxes, whereas cognita do not. We note then that the integrity of a city or like gathering of population does not depend on its having a wall round it; still less does its existence. Certain economic, social, etc., causes or 'forces' have operated to bring about the local condensation; and so long as these continue to operate, the condensation will remain. Similarly, there is no need to postulate anything in the nature of a mind-itself to contain or hold together the population of cognita that forms it; forces, suitably specified, acting between the cognita themselves are all that is required. Again, the individuality—i.e. the characteristic culture, etc.—of a city may be maintained over indefinite periods, despite considerable changes of population, though every immigrant or emigrant will alter its character to some extent.[1] This is an important point as regards the preservation of individual identity in the case of the mind, and it is, of course, again necessary to avoid reasoning that because Birmingham would still be Birmingham even if you exchanged the entire population with that of Edinburgh (Birmingham being tacitly defined for this purpose as the city in such and such a geographical position), therefore there is something (the mind-itself) which would remain the same, even if the whole of its content were altered.

I will now leave these illustrative images on one side for the

[1] Cf. Hume: 'I cannot compare the soul more properly to anything than to a republic or commonwealth, in which the several members are united by the reciprocal ties of government and subordination, and give rise to other persons who propagate the same republic in the incessant changes of its parts' (*Treatise of Human Nature*, Bk. I, Pt. IV, Sect. 6; Everyman Edition, Vol. I, p. 247). (H. H. P.)

time being, and revert to a more direct discussion of the mind as I understand it.

64. *Consciousness.* I think the reader may fairly object at this point that though I may have given a sufficiently clear, if necessarily imperfect, indication of the way in which I conceive cognita to be grouped and concentrated so as to form minds, and of the fact that in any mind there is an incessant coming and going and flux of the constitutive cognitum groups (as the midges in the cloud, or the citizens in the city), I have yet omitted the most important thing of all, namely, 'consciousness'. And to speak of a mind without consciousness is like discussing *Hamlet* with no reference to the Prince of Denmark. So far I agree. Further, it may be urged, there must be *something*—call it what we will—which *is conscious of* the groupings and sequences of cognita as they come and go in the field of consciousness. But here I am radically and immovably intransigent. In no circumstances whatever will I allow the introduction of any inherently unobservable entity; for to do so would automatically vitiate all further discussion, rendering it futile and abortive from the start, inasmuch as no assertion or denial could ever be verified or disproved.

According to me, then, a so-called individual mind[1] is a certain kind of grouping or concentration or, as I have spoken of it above, a condensation of cognita, having no set limits, but held together in greater or lesser closeness by forces acting between them; and I contend that consciousness, or 'being conscious', is not a property of some Ego or other metaphysical (non-existent) entity extraneous to those cognita but *is* that system of forces operative between them—in much the same sense that we might say that gravitation *is* the system of forces operating between material masses. That is to say, if there are no cognita there is no consciousness—which is, 'rightly', equivalent to denying the possibility of being conscious of nothing—but wherever two or three cognita are

[1] Note in passing what we shall have occasion to emphasize later, namely, that the word 'individual' is actually a serious misnomer, for the human mind seems one of the most highly divisible things we know!

athered together, so to say, there is consciousness in the
nidst of them.[1] Broadly and very provisionally speaking, I
onceive that the degree or intensity of the consciousness
haracterizing any aggregate or group of cognita will depend
n one or other or both of two factors, viz. the number of
ognita in the aggregate and the intensity of the forces acting
etween them, though I should be sorry to be called upon to
ive anything approaching a formula connecting the quanti-
ies involved. But the conjecture seems fairly safe, since I
annot see any other factors on which it could conceivably
lepend. Metaphysical unobservables being barred, and
ssuming that it is reasonable to speak of consciousness vary-
ng in degree at all (as common experience seems to make
lain that it is), then any such variations must be correlated
vith some feature or features to be found within, so to say,
he system of cognita concerned; that is to say, either with the
ntrinsic character of the cognita (whether red, blue, visual,
uditory, etc.) or their number, or the relations (forces, etc.)
etween them. The first seems to me to be implausible, unless
ve count 'vividness' or some such quality, as an intrinsic
haracteristic; but I should have said that vividness was much
nore in the nature of a measure than a cause of variations of
ntensity or degree of consciousness. The whole question is
dmittedly obscure, but not of primary importance here.

The point is that we must *not* drag in metaphysical entities,
uch as *Egos*, and say that they are 'conscious of' the cognita,
nd we therefore *must* locate the phenomenon of consciousness, or 'being conscious' within the cognitum systems them-
elves. There seems, however, to be one possible variant of
he above account, and this I will discuss when I have dealt
vith the problems of the 'empirical self' and 'self-consciousness'.[2]

65. *The Empirical Self.* Although all metaphysical entities,
uch as the *Ego*, must, like the *Ding*, be totally rejected from
our speech and thought, it is none the less perfectly legitimate,

[1] This involves implications which are very liable to affright the faint-
earted and will be discussed briefly below, but I do not think they will
rove as formidable as they may appear at first sight.

[2] Cf. Sect. 71, below, pp. 177–180.

and in practice convenient to the point of necessity, to use the term 'self' to refer to whatever it is that is characteristic of one person but not of another, though this is rough and somewhat anticipates the discussion.

There seems to me to be two classes into which the cognitum groups to be considered in this connexion may properly be divided: first, those which would ordinarily be called sensa (or images) of endosomatic origin; second, those memory images arising out of (I speak rather colloquially) our personal life history; and in each case the words 'in so far as they are peculiar to ourselves' should, I think, be added, for we are at the moment interested in what differentiates one mind from another and gives it what we call its individuality rather than in what makes all minds to some extent alike.

What we have termed the field of consciousness of any mind at any moment commonly includes, albeit marginally, a certain proportion of cognita (sensa) of somatic origin—notably endosomatic—forming what Broad, I think, has called 'that vague mass of bodily feelings' which is an invariable accompaniment of more overt experience. This mass of feelings changes in details, of course, from moment to moment, and considerably from year to year, but it none the less forms a kind of semi-continuous nucleus, if I may use the phrase, round which our experiences of the outer world are clustered—rather like a rope of which no single fibre is more than a few inches long, but none the less is continuous when considered as a whole. But one may reasonably suppose that this core of bodily feeling is for the most part, far from uniquely characteristic; the vague awareness of breathing and heart-beat, of muscular movement, of visceral happenings, etc., must be substantially similar in all normal people, and it will only be special pains, hungers, injuries, illnesses, and the like which will at all sharply differentiate one such set of cognita from another.

Environmental experiences or life histories, on the other hand, differ very widely, and it is, I think, in our memories of these (using the term in the widest sense, of course, to include vicarious experience by reading, etc.) that the main differences are to be found. I should say that the difference between Whately Carington's self and that of Tom Jones is much more a matter of the memory images of past experience having potential access to W. W. C.'s field of consciousness, but not to T. J.'s, and vice versa, than of differences in the present sensa and memory images of their bodily feelings.

The accessibility of these memories, or, more accurately, the probability of their making an appearance in the field of consciousness is, however, also highly variable. Recent experiences have in general a relatively high probability of doing so, and more remote experiences a lower; but there are many and striking exceptions to this rule. Moreover, speaking for myself at least, there are whole tracts of my life which seem to me to have faded completely into oblivescence and are as if they had never been, though doubtless the experiences concerned still exert an influence on what I am and how I behave and think to-day.

It is not easy, but fortunately not important either—at least for our present purpose—to decide whether the word 'self' should be used as virtually synonymous with 'mind' to include all the cognitum groups of psychological or mental type which have this relation of potential access to the field of consciousness, and which in turn serve to define and identify it, or whether it should be restricted to those which remain, or would remain, after counting out all which are of what we would normally call obviously external origin.[1] Personally, I see no necessity for speaking of the 'self' at all, except colloquially or when we are specifically discussing 'self-consciousness'. If we believe ourselves to be entitled to speak of anything other than the body, then it seems to me that the word 'mind' will sufficiently serve all purposes, provided it be suitably qualified or delimited, if need be, to suit the particular occasion.

But in any event this is no more than a point of technical and linguistic convenience, and does not in any way affect the

[1] This reservation seems desirable, lest we find ourselves inadvertently including as part of the 'self' groups forming parts of material objects contemporaneously perceived, which would be implausible.

main and all-important point at issue, namely, that whether we choose to use the word 'self' or not, and whatever we may refer to by it when we do, there is no warrant whatever—but very much the reverse—for conceiving of it as a metaphysical entity endowed with the magical property of being 'conscious of' this or that. The 'self', like everything else, must be empirically observable, or else ruled out as non-existent.

66. *Further Discussion of the Mind.* I am not at all satisfied that my account of the mind up to this point will have been sufficiently clear to carry conviction, and I think it may be profitable to cover much the same ground again, in part at least, from a slightly different angle and using slightly different language, at the same time introducing some discussion of the 'forces' which I suppose to be operative.

The word 'cognitum' is somewhat arid, and 'cognitum-group' a trifle clumsy for general use, especially if one has to be constantly adding some qualifying remark to the effect of 'organized according to psychological laws' or the like. I have found it helpful in the past to use the term 'psychon' to denote such cognitum groups,[1] and to speak of the mind as a 'psychon system', and I shall adopt this practice henceforward here, whenever convenient. It must, however, be most clearly understood that, by introducing a fresh word, I am not thereby tacitly postulating the existence of a new entity, least of all a metaphysical entity. Cognita are still cognita whether they form parts of minds or parts of material objects, just as men are still men whether we call them 'financiers' when they are organized in cartels, or 'convicts' when they are subsequently breaking stones on Dartmoor. A psychon may, in principle, be a single cognitum—e.g. an undifferentiated colour expense—but in general it will be, so to speak, more analogous to a molecule than to an atom; that is to say, it will in general consist of two or more cognita more or less stably conjoined, e.g. any coloured patch has both a colour component and a shape component, each of which must either be a cognitum or resolvable into cognita. It is rather a nice point whether the oval yellow patch which I am said to

[1] Cf. *Telepathy* (Methuen, London, 1945), pp. 96 ff.

cognize when I see a lemon is to be called a psychon, since it is undoubtedly a part of my field of consciousness and therefore of my mind, or a 'hylon' (to coin another word for a moment which I shall not use again), on the ground that it is also indubitably part of a material object. Actually, of course, it is both, and there is no reason why it should not be. But a memory image of it would unquestionably be a psychon. But the point is only a triviality.

According to the view I am here developing, the mind is a kind of witches' cauldron of psychons, to which new ingredients are perpetually being added in the form of, as we say, new experiences, and perpetually throwing up fresh configurations of assorted bits and pieces to the surface (field of consciousness)—with the all-important reservation, of course, that there is no cauldron, and (so far as I know) no witches. The forces between the psychons are here represented by the differing size and density of the various bits and pieces in the brew, and by the convection currents, etc., in the fluid surrounding them. Some of the pieces are so light that they remain almost continuously at the surface, others of medium density are easily thrown up, so that the probability of their appearing on the surface in any given period of time is considerable, while others are so large and heavy that nothing short of a major upheaval will bring them to the top. Note, in passing, the introduction here of the notion of the *probability* of any given item appearing at the surface; this could be used as a measure of the relative densities of the items, or different types thereof, and therefore of the 'forces' acting on them, and it is, I think, the only quantitative measure we can apply in the case of the mind. The very light bits represent the psychons (cognitum groups) which are almost always present in the field of consciousness—somatic sensation and the like; those of medium density are the ideas and memories, etc., which we can easily recall and frequently think of; while the very heavy items are the ideas, etc., which we describe as being repressed into the subconscious or unconscious.

But the analogy is inevitably imperfect, and there is another which in some respects I like better, as tending to bring out the notion of forces acting between the psychons, as opposed to convection-currents, etc., acting on them. There is a very beautiful experiment, which I have not seen for very many years; it used to be demonstrated at one time, to present a kind of model of the way in which electrons were at that time supposed to arrange themselves inside the atom. You take a dish of mercury, and float on it one or more steel balls, each of which is magnetized in the same way as the earth, namely, with one north and one south pole. Above or below the surface of the mercury you arrange one pole of a bar magnet. If you then float a single ball on the surface it will take up a position vertically below (or above) the end of the magnet, with its north pole up, of course, if the south pole of the magnet is the nearer to the mercury, or vice versa. If you add a second ball, the two will arrange themselves at equal distances on each side of the point immediately below the magnet (there is nothing much else they could do). Three balls will not form a row, with one in the middle and one on each side, but an equilateral triangle with its centre below the pole, and this again is what I think one would expect. But four balls do not form a square; they arrange themselves as an equilateral triangle with the fourth ball at the centre. Five balls do the same thing, but with two at the centre. Six balls do not form two triangles, but a regular hexagon, and a seventh will take up a position at the centre of this.

If we imagine the balls to be of indefinitely numerous shapes and kinds, resembling each other only to the extent that forces may act between them, blur the outlines to taste, and then translate the whole thing into as many dimensions as we feel disposed, we shall have a fair picture—within its limits—of the mind as I conceive it. In particular it suggests the kind of way in which the pattern or configuration of relationships between the cognita which we should usually speak of as the content of the mind, may be altered by the intrusion of fresh items, and yet be largely determined by the whole of the previous content. But the picture is in certain respects misleading, as all such pictures or models are bound to be; it would, for example, be illegitimate to argue that because, in this illustration, we postulate a magnetic pole above the surface of the mercury, therefore there must be an organizing 'somewhat' (ego, etc.) which determines the pattern of the cognita forming the mind.

67. *Forces in the Mind.* Since we may not introduce inherent unobservables (equivalent of magnetic poles, etc.) into our account, we must rely solely on 'forces' acting between the cognita themselves; and it may be wise to remind ourselves that a 'force' is not something directly given in experience in the sense that objects and their movements may condensedly be said to be given. Our experience of sensations of pressure, etc., often leads us to suppose that it is, but this is not the case; all that is directly cognized are the cognitum sequences described as objects and their movements. Force is a derivative conception of a fictional and strictly tautological character. When we say with Newton that a (material) body continues at rest or in a state of uniform motion except in so far as it is acted upon by an external impressed force, we are actually saying no more than that a body continues in a state of uniform motion (of which the magnitude may happen to be zero), except in so far as it doesn't. If we ask why a body has changed its rate of motion, we are told that it is because it has been acted on by a force; and if we ask how it is known that a force has acted on it, the answer is that it is because it has changed its rate of motion. There is nothing wrong with this; it is the only way of avoiding the introduction of unobservables, which may be related to what Newton had in mind when he said *Hypotheses non fingo.*

Somewhat similarly, when we ask what *sort* of force it is that acts, we are obliged to indulge an analogous tautological activity. We observe that material bodies describe (*in vacuo*) a parabolic path, and that planets move round the sun in ellipses, having the sun at one focus (or very nearly so), and —to put it colloquially—we fake up a 'law of force' of such a kind as will give these results. If the observations were different we should have to fake up a different 'law of force' to suit these facts. The observations come first, the 'laws' merely embody them.

In the case of the mind we must necessarily follow the same type of procedure; that is to say, we must observe the types of regularity which do in fact characterize the flux and succession of cognitum groups (psychons), and then devise some sort of a 'law of force' to cater for them.[1]

Now it seems clear to me that there are two, and probably only two, main types of phenomena which take place in the mind, or in mind generally. There are phenomena of Association and phenomena of Dissociation; and to cater for them we need either two laws of 'force', one being of 'attraction' and the other of 'repulsion', or a single law which will give the one result or the other according to circumstances which we shall have to specify. Let me hasten to add, lest false hopes arise in the reader, that I am not in a position to formulate the necessary laws here—or indeed anywhere else. Before we can do so we must first decide what are the fundamental quantities (or equivalent conceptions) in terms of which they must be formulated—i.e. whatever it may be that, in the realm of mental phenomena, will take the place of Mass, Distance, and Time in that of physical phenomena—and I doubt whether we are yet capable of doing even so much as this in the present undeveloped state of the subject. Second, we need a mathematician of high calibre (which I most emphatically am not) who will study the facts and excogitate a formula to cover them, expressed in terms of these 'quantities'. But these are purely technical problems which will doubtless be solved in due course, and do not affect the main issue considered here. None the less, it seems possible to put forward a few reflections which may serve in some measure to clarify our conceptions of the kind of thing that seems to be going on.

[1] Strictly speaking, the term 'force' is a misnomer in this connexion, or presumably so. Force has the dimensions MLT^2, i.e. it is a mass multiplied by an acceleration; and the word is not properly applicable to anything else. It seems to me unlikely that cognita can properly be said to have mass, to exist in space, or to be subject to acceleration, so that the terms used in describing their comings and goings are unlikely to correspond exactly to those used in dealing with material objects. But the provisional use of the term 'force' seems legitimate, provided we do not lose sight of its analogical status.

68. *Forces in the Mind* (2): *Association.*[1] We are all of us more or less familiar with the phenomena of association—'the association of ideas'—as it is commonly phrased. Putting it colloquially, the sight or thought of Bread is more likely to suggest Butter or Jam than Turpentine or Boot-blacking; Bacon tends to suggest Eggs, Bryant to suggest May—and so forth in innumerable instances. This is simply because we have experienced the words or the objects in conjunction, and they have become, as we say, 'associated', so that re-experiencing one member of the pair tends to recall images, etc., of the other.[2]

This is usually formulated in some such way as by saying, 'If two or more objects, A, B, C, . . . etc., or "ideas", "images", etc., A′, B′, C′, . . . etc., of those objects, are present to consciousness (or in the field of consciousness) together, or in close succession, then re-presentation or recurrence of any one of them is liable to recall the others' (or, preferably, '. . . then the probability of one or more of the others recurring within any given period thereafter is greater than if they had not previously been presented together'). In the terminology used here we shall, of course, say 'cognitum groups' rather than 'objects', 'ideas', 'images', etc. It would perhaps be safer to say 'psychons', on the ground that although the probability of the cognitum groups constituting a material fried egg following (in their proper pattern) the re-presentation of those constituting a material slice of bacon is greater than if no such re-presentation of the bacon cognita occurred, yet this probability may be small, despite previous conjunctions, compared with the recurrence of groups constituting a purely imaginary fried egg. But the point is evidently trivial.

The essence of the matter is that cognitum groups, notably those known as psychons, once they have been 'in consciousness', as we say, in conjunction (or close succession) do tend to hang together, so that recurrence of one cognitum, or

[1] Cf. the author's *Telepathy*, Ch. VI, for a fuller treatment. (H. H. P.)

[2] Note that this is, to all intents and purposes, identical with the account we gave of how words acquire meaning.

group (or one closely resembling it[1]) does tend to recall or facilitate the recurrence of the others.

There are, of course, 'sub-laws' of association; e.g. that the more often the conjunction of A, B, . . . etc., takes place, the greater the probability of B (or A) accompanying or following the re-presentation of A (or B); or that this probability is greater as the occasion of initial presentation together is more recent, etc.[2] But I do not think we need concern ourselves seriously with these sub-laws here, beyond pointing out that, given the basic fact of association, they seem natural enough on the kind of view of the mind we have been considering. Given that 'compresence' (to use the technical term) establishes, as it were, a kind of link between the relevant psychons, then it is easy to understand that repeated compresence can reasonably be supposed to establish many links (equivalent to 'a stronger' link) between A and B, say, so that the one is thereby rendered more likely to follow the other, etc.; and that the compresence of A with a number of different psychons B, C, D, E, . . . etc., such as may be supposed to occur in the course of assorted experience in the course of time, would reduce the probability of any given one of them (C, say) following the re-presentation of A. Roughly speaking, the more recently we have dropped a given bunch of items into the witches' brew, the more likely they are to come to the top

[1] Presumably any two cognitum groups said to resemble each other can in principle be analysed into constituent elements of which some would be said to be different from each other, but others identically similar. If so, then it is meaningless to say that two members of any class of the latter category (e.g. two identically similar reds) are numerically distinct, unless they are cognized simultaneously. The point is suggestive, but I do not think we need follow it up here.

[2] For those who enjoy the wilder flights of speculation it may prove diverting to indulge the fancy that what we call the 'properties of matter', i.e. the observed high probability (it is never certainty) that cognitum groups of one type, e.g. tactile, will closely follow cognitum groups of another type, e.g. visual, or vice versa, in the case of any material object, may have ultimately an associative origin. But I am not to be taken as in any way endorsing this suggestion at the present time. [It would follow that the laws of Psychology would after all be more fundamental than the laws of Physics: and Idealism (though of a shockingly untraditional kind) would be nearer the truth than Materialism (cf. Ch. IV, Sect. 56, pp. 137–140, above). (H. H. P.)]

together—there will be less chance of their having been separated in the turmoil of the boiling broth; and the more often we drop in a bunch of a certain specified constitution, the greater the chance (within the limits of the analogy) of any two or more members of such a group coming to the surface together.[1]

To revert: I am more interested here in the phenomenon of association as indicating (as we would ordinarily say) the operation of forces acting between the cognita (psychons) constituting a mind (or Mind generally) than in the particular form it takes; and it seems to me that it requires us to think of a very odd sort of force, markedly different, in certain respects, from those to which we are accustomed to deal with in the physical world.

We are familiar enough with the fact that two material particles or bodies 'attract' each other, and with saying that they do so with a 'force' which varies directly as the product of the masses and inversely as the square of the distance between them, and *mutatis mutandis* in the case of electrically charged bodies. But, although there is something definitely

[1] I hope I need hardly emphasize here that I am talking throughout in terms of groups or aggregates of psychons; that is to say, substantially in 'gestalt' terms, and not in terms of the old 'hook-and-eye', single-track, one-dimensional associationism. This not only did not work, but was so manifestly absurd as to discredit the very word 'association' to a point at which many psychologists seemed almost to deny the occurrence of the phenomenon. The suggestion that the idea A inevitably calls us 'by association' the idea B, that B then calls up C, then C, D, and so on—like trucks hitched to a locomotive—is so absurd that I must confess I thought of the mind in terms of associational phenomena for many years before realizing that anyone had been so imprudent as seriously to entertain it. The whole essence of the business is that ideas, psychons, cognitum groups, or whatever you please to call them, are linked, each with the others, in aggregates, systems, or *gestalten*, so that the recurrence of any one *tends* (in varying degree, of course) to promote the recurrence of any other member of that group. Nor do I see any reason for setting limits to the groups. Every psychon said to be 'in' a mind is linked, as part of an aggregate, directly or indirectly and more or less closely, with every other. The links, of course, are metaphorical, and when we say that the link between A and B is stronger than that between A and C, or the like, this is only a convenient way of saying that, other things being equal, the probability of A being followed or accompanied by B is greater than that of it being followed or accompanied by C. [Cf. Sect. 78, p. 202, below. (H. H. P.)]

reminiscent of gravitation about the way in which psychons tend to cluster in groups, there is this very queer difference, that they only do so *after* they have been brought or have just 'come' into conjunction. This is more suggestive of the way in which two pith balls, say, will repel each other after a charge initially concentrated on one of them has been shared between them—except that there is something like attraction and not repulsion here. Moreover, I do not see anything to correspond in the case of association phenomena with either the inertia of material bodies (at least I think not) or with the movement which alone prevents satellites, for example, crashing into their planets, or planets into their sun.

Evidently we must be extremely cautious about attempting to formulate the 'laws of force' governing the changes in cognitum (psychon) systems merely by arguing by analogy from those operative in material systems. And, as I have already indicated, we must in no circumstances allow ourselves to think of 'a force' as something capable of existing independently of the entities on which it is said to operate. All we know are cognitum groups and sequences, and any 'law of associational force' must be in the nature of a compendious generalization of the way in which these change and succeed each other, just as any 'law of gravitational force' must be a compendious generalization of the way in which material bodies (in certain circumstances) move with respect to each other.

69. *Forces in the Mind* (3)*: Dissociation.* The tendency of one idea (psychon) to call up, as we say, another, and for psychons to cohere or stick together in clumps, and so forth, is perhaps the most familiar and noticeable feature of mental activity. Every hour of the day something or other 'reminds us' of something else, while every writer, I imagine, must have noticed the way in which one of a number of possible words tends to come to mind merely by virtue of assonance with some other word just used.

But it is scarcely less noticeable that some of these clumps or clusters of ideas seem much less easily evoked than others, or even to be actively incompatible with those which most usually occupy our minds. Like association, this tendency to dissociation seems to operate, *en gros et en detail*, in every phase of mental life, though naturally in very varying degree and in respect to innumerable planes of cleavage, as we might call them. At one end of the scale we have the changing moods of everyday experience, which seem to call for no special machinery to explain them, or the fact that a man preoccupied with his professional activities may present a personality very different from that which he shows when absorbed by a private hobby. But it seems impossible to draw a hard-and-fast line between such quite normal fluctuations as these and the extreme cases of multiple personality familiar to psychopathologists.[1] Or again, we most of us know what it is to forget, temporarily, some name or fact which we should ordinarily expect to remember, and it is often possible to trace an associative connexion between the forgotten word, etc., and some person or incident definitely repugnant to us; and psycho-analysts assure us, on excellent evidence, that there are, in most people at least, certain systems of psychons (complexes) so violently repugnant to us that they cannot come to consciousness at all, but remain for ever 'repressed' in the subconscious, unless special steps are taken to bring them to the surface.

I am not at all concerned here to discuss these phenomena in detail—there are plenty of technical works which do so far better than I could—but merely to emphasize their occurrence, and with it this aforementioned tendency for psychon systems to show varying degrees of incompatibility with each other. I must also emphasize (as indeed I have remarked elsewhere) that all our studies of these mutations of personality etc., make it very clear that the mind is far from being the serene and sovereign state we like to picture it, but is much

[1] Such as that of the Reverend T. C. Hanna, Ansel Bourne, the various Miss Beauchamps, the Doris Fischer case, and many others. The most valuable short discussion known to me is Dr. T. W. Mitchell's paper, 'Some types of Multiple Personality', in *Proceedings* of the Society for Psychical Research, Vol. XXVI (1912–13), pp. 257–85. [Cf. also William James, *Principles of Psychology*, Vol. I, Ch. X, pp. 371 ff. (H. H. P.)]

more like a loose federation of semi-autonomous republics all too often seriously in conflict with each other.

The question then arises whether it is necessary to attribute these dissociative phenomena to the operation of a special 'force' *sui generis* tending to make certain systems fly apart, as it were, or whether we are to regard the non-appearance of some of them as merely a matter of default, if I may put it so. By the last phrase I mean something like this: what we actually *observe* is that certain systems—for example, our very natural and indeed inevitable desire to eliminate our father-rival and possess ourselves exclusively of our all-important mother—do not in fact often, if ever, present themselves overtly in consciousness; that is to say, the probability of their doing so is very small, though they unquestionably make their presence felt in various indirect ways. But this might be merely because the probability of other systems presenting themselves is very large; to put it slightly differently, the allegedly 'repressed' systems might merely be crowded out by the claims of other systems, not actively repelled by them. I think, however, that everyone who has any considerable knowledge of these matters would agree with me that this view is quite untenable; or at any rate that, if it is to be held at all, we must find means of attributing a quite extraordinary positivity to mere negation! These repressed complexes often appear to possess a remarkable degree of vigour; they seem to writhe and struggle for expression for all the world as if they had a conscious life of their own (as I think they probably have);[1] and it is only when at last, and by one means or another, they are brought to the surface and related in consciousness to the other constituents of the personality that they burst like bubbles and lose their potency. This is, of course, in a high degree anthropomorphic and metaphorical, but it should serve to indicate the kind of impression that is derived from study of the relevant facts.

We find ourselves, then, in the rather difficult position of being obliged to say that cognitum groups may in certain circumstances behave as if they attracted each other, so that

[1] Cf. Sect. 64, pp. 158–9, above. (H. H. P.)

they cohere in more or less stable clusters or constellations, but in others as if they repelled each other and actively eschewed each other's company.[1] It would be hardly decent to keep the two sets of phenomena, associative and dissociative, in watertight compartments, and excogitate one 'law of force' for the one set and another for the other; yet, if we do not, it looks as if we shall have to work out a kind of Janus-faced law capable of yielding, so to say, attraction and repulsion from the same formula.

Fortunately the situation is not without precedent, though actual solution of the problem is for the probably far-distant future.

Newton's law of gravitation enabled us to account, almost exactly, for the observed movements of the planets of the solar system, and for a long time was held to be the last word on the subject. It was concerned solely with how the attraction between material bodies varies with their mass and the distance between them. But it did not explain even the movements of the planets quite exactly; the movement of the perihelion of Mercury, in particular, remained recalcitrant. Later, other considerations, notably those arising out of the Michelson-Morley experiment, led Einstein to his famous theory of relativity. This proposed a law of gravitation not quite identical with that of Newton, though leading to substantially identical results for most practical purposes. But it enabled us to clear up the observed irregularities of Mercury, and it was remarkable, for our purpose, for including a term of repulsion as well as of attraction.[2] What this came to was that bodies attracted each other, substantially as Newton maintained, so long as the distance between them was not too great, i.e. within the limits of the solar system, or even of

[1] In the case of primitive instincts in conflict with social conventions, the problem is fairly easily dealt with in terms of conditioned reflexes, nerve paths, synaptic resistances, inhibitions, and the rest of it; but it is very much more difficult when we are concerned with abstract conceptions, which may be just as incompatible as different modes of behaviour. Besides, I suspect that to talk in these terms is tacitly to beg the whole question of the matter-mind relationship which we are discussing.

[2] For a semi-technical account, see Eddington, *The Expanding Universe*, or *New Pathways in Science* (Cambridge University Press), Ch. X.

a galaxy; but for really large distances (by astronomical standards) the effect was reversed; the attractive term became less important then the repulsive, so that at these distances bodies repel each other instead of attracting. And when astronomers came to observe the motions of the galaxies outside our own (spiral nebulae) they found that they were in fact receding, at a rate proportional to their distance from us, in the way the theory predicted; that is to say, they were behaving as if they repelled each other. The analogy is not at all close, except in its broadest structural outlines. But it does serve to show that to devise a 'law of force' or equivalent[1] which shall cater for both attractive (associative) and repulsive (dissociative) phenomena, is not at all such a contradiction in terms as might appear at first sight.

It may well be that when mathematical psychologists of the future tackle the phenomena of mind as seriously as mathematical physicists of the past have tackled the phenomena of matter they will find it not too difficult—working, of course, in categories quite other than the material ones of mass, time, and spatial distance—to devise formulae which will bring both association and dissociation within the scope of a single unitary theory.

Just how it will work out is not important, and I have written at this length on the point mainly because I want to indicate that there is no justification for supposing, as many do, that the phenomena of mind are necessarily beyond the scope of the methods which have been so successfully applied to those of matter. What is important is to remember that the mind, like a material object, is and can only be a system of cognita, and cognitum groups and sequences, however different may be the regularities of pattern in the two cases; and to refrain at all costs from dragging in metaphysical unobservables to 'explain' (*obscurum per obscurius!*) phenomena which at present seem to us strange and even chaotic.

70. *Conflict and Equilibrium in the Mind.* I do not at all know whether I have succeeded in conveying anything like the

[1] I say 'or equivalent' because, technically speaking, relativity theory does not use the conception of 'force' but talks about 'curvature of space'.

picture of the mind which I wish to convey. Since all pictures, like models, are bound to be in greater or less degree misleading, and words are imperfect vehicles of expression, it seems probable that I have not; still, perhaps the impression I have given will serve well enough for a basis of further discussion.

We have envisaged the mind as a system of cognitum groups, which we have called 'psychons' for convenience, organized in a particular way, namely, according to the 'laws of psychology' of which at present we know relatively little. But we do know, first, that psychons which have been in a certain relation to each other, described as 'present together' or 'in close succession in consciousness', tend to cohere in sub-systems; and, second, that there is something akin to repulsion, which may be called incompatibility, such that some of these sub-systems tend to be driven, as it were, apart, so that they seldom, if ever, occupy a field of consciousness together. I suggest that as a kind of overriding or summarizing 'law' of mental phenomena we might entertain the hypothesis that the mind works according to a Principle of Least Conflict (somewhat analogous to the principle of Least Action in physics), so that mental events take place (i.e. cognitum groups adjust their relations to each other) in such a way as to reduce to a minimum at any moment the 'conflict' resulting from the incompatibility of some with others.

I should feel that this was a good deal more impressive than it is, if only I were in a position to define 'conflict' more precisely than by saying that we all know what it feels like. The condition is familiar enough in cases where two or more possible courses of overt action present themselves, and we cannot, as we say, 'make up our minds' which to take; but it equally obtains in even the most abstract thinking, when we hesitate between alternative forms of words or other symbols till we find the one which best satisfies all requirements, or at least leaves us with the least sense of dissatisfaction. And the same principle applies, I think, throughout; and the mind chooses always the path of least resistance, little though we may realize it. More accurately, the psychon systems automatically take up whatever configuration involves the minimum of strain between them—though even this is vague and figurative. The martyr elects to be thrown to the lions rather than drop a pinch of incense on the altar of Jupiter because, distressing as this decision doubtless is, it is less so than renouncing his Christian principles; the neurotic develops a functional disorder because this is less trouble than facing the exigencies of normal life; the dreamer indulges a symbolic dream because this, though not wholly satisfying his repressed desires, gives rise to less conflict than either envisaging them openly or allowing them no expression at all; the mathematician writes down correctly (if he is lucky) the next step in the working of his problem because this and no other least conflicts with the sum total of his relevant knowledge; the poet seizes on the perfect phrase which springs into his mind as the resultant of pressure by all those verbal and aesthetic sub-systems to some of which any other phrase would be offensive.

All this is considerably anthropomorphic and metaphorical, and is not intended to be too literally interpreted. The point I wish to suggest, despite some of the above wording, is that there is not involved any metaphysical entity (the Ego, the Mind-itself, etc.) which chooses, selects, and decides; but that the content of consciousness, as we call it, at any moment (except in so far as it is imposed from without) is the resultant of the interplay of associative and dissociative forces operating between the cognitum groups which form it. This sounds, and is intended to sound, severely mechanical; but I shall deal below, to some extent at least, with the hollowness of the objections which many people will feel towards it, and particularly with the problem of Free Will.[1]

I think it important that the mind should be conceived as a system perpetually changing the configuration of its constituents under the influence of forces acting between them,[2]

[1] See Sect. 75, below, pp. 190–8 (H. H. P.),

[2] Let me say again that 'force' is not to be thought of as something capable of existing apart from the entities 'on' which it is said to operate. In due course we shall probably abandon the interim conception of 'force', as the physicists have done, in favour of something equivalent to the 'curvature of space' in which they now prefer to discuss fundamentals.

in a manner we might figuratively speak of as 'seeking' equilibrium. But it is not less important to remember that there is nothing static about it, and that although the configuration, as I have called it, taken up at any moment is one of minimum strain or conflict, this does not mean that it is necessarily or ever one of zero conflict.

71. *Consciousness, contd.* We may now go back to the question of Consciousness, which is obviously of central importance. I need hardly say that the subject is one which bristles with difficulties, not all of which would I for a moment profess to be able to solve at the present time. The all-important point is that *unless* you are prepared to drag in metaphysical unobservables (in which case, as I have so often insisted, you are foredoomed never to know whether your statements are true or false, and are only wasting time in making them) you *must* give your account of consciousness in *some* terms of cognita, or cognitum groups and sequences, or the forces acting between them, and of nothing else whatever; and we have seen that 'forces' are not independent existents. This seems to me as certain as anything can well be, but it would be idle to deny that it raises problems of its own such as do not arise if we basely evade the issues by postulating inherently unobservable 'souls' or the like.

To start with: I have committed myself to the view that whenever two or more cognita are linked together (or words to that effect), there will be some degree of consciousness between them; and I think I must stick to this, though there may be a possible way out which I shall note in a moment. I have also said that a material object (such as a rock) consists of cognita linked in a certain type of pattern which we describe as conforming to the laws of (material) physics. Are we then to say that a rock or other material object is conscious? I think we must, but I do not think that this statement is at all so alarming as it may appear at first sight. It is all a matter of degree, and I think (though I may be speaking somewhat loosely) of the amount of associated experiences—or something very like this.

If we look at the matter biologically, and trace consciousness down, so to say, from such highly intelligent and acutely conscious human beings as the reader and myself, through other less brilliant humans, through the higher animals, and the lower, down to the humble protozoa, we shall find it as difficult to say just where consciousness leaves off as to say where 'life' leave off; and if we cannot devise and justify a criterion for deciding where it leaves off, it seems arbitrary and illogical to say that it does. But the experience, direct and vicarious,[1] of an amoeba, or of a single cell in my body, or even of my unhappy liver, is infinitesimal compared with what would ordinarily be called 'my' experience—point several zeros one per cent perhaps—and I see no particular objection to allowing these lowly organisms a corresponding degree of consciousness if logic seems to require it.

As for what we commonly describe as inanimate objects, the position is perhaps a trifle more difficult, though I still have no great objection. After all, as Ruskin, I think, pointed out, 'a rock is a darn' good sitter' (or words to this effect); and if my own experience were limited to sitting, or rather to just being in the same place, without any sense-organs to enlarge it, I should, I conceive, show no more signs of consciousness than a rock. This is admittedly rather rubbish, and is not to be taken seriously; but I do want to emphasize the point that, if the view I have put forward does logically involve attributing some kind or degree of consciousness to objects not usually credited with it, we are not necessarily involved in anything very serious.

We might, however, get out of this difficulty, if we wish to, by saying that consciousness only obtains between such cognitum groups as are in fact organized in those patterns which conform to the laws of psychology, i.e. in minds or as mental phenomena. This seems an easy way out, and will probably appeal to many; but personally I do not greatly care for it. It seems to me to come rather near begging the

[1] It is not easy to see how an amoeba can have 'vicarious experience'—unless it possesses a language or something equivalent, enabling it to receive information from its fellow-amoebas. (H. H. P.)

question, and further analysis might well show that it actually does so; moreover, it might lead to misinterpreting, on account of *a priori* prepossessions, instances of quasi-conscious behaviour on the part of (for example) physiological structures, such as could be better understood if we admitted the possibility of their possessing some kind or degree of consciousness. On the whole, I should prefer to suspend judgement, with a slight bias towards the virtual universality of consciousness, till further study pushes us one way or the other.

But it is, above all, important to remember that, if we find ourselves adopting this more catholic view, we are not to suppose that 'consciousness' necessarily implies the same kind or degree of state with which we are ourselves familiar.

Another difficulty, though I think it trivial, arises as follows: in an earlier account of what I termed the 'field' theory of consciousness, substantially the same as this, I was at pains to emphasize that consciousness *per se* was not a unidirectional or unilateral affair, in which one entity (the Ego, etc.) was conscious 'of' another, but a matter of the relationships of tension or the like obtaining between what I should now call cognitum groups, etc. Using the analogy of gravitation to illustrate my point, I observed that, whereas we say 'The sun pulls the earth', we know that the earth pulls the sun just as hard, so that 'gravitation', which I was likening to consciousness, was a state of affairs between the two to which both were necessary. This was taken by a critic of distinction as implying that, on my theory, the statement 'I am hearing a loud bang' was to be regarded as equivalent to the statement 'A loud bang is hearing me'—which, of course, is absurd. I do not, I need hardly say, wish to imply anything so ridiculous as this. To say 'I hear a bang', or 'I am conscious of a bangy noise', does to be sure imply that bang-cognita are compresent in a certain field of consciousness with those others described as 'I' at the moment concerned, and that if one set or the other be lacking, there is no consciousness of the bang by me. But to say that each of two terms is necessary to a certain kind of relationship between them is not, so far as I can see, to say that that relationship is symmetrical. In any event, to affirm or deny that the bang hears me is, as nearly as possible, meaningless, since there seems no possible way of determining whether it does 'hear' me or not.[1] In these circumstances, the difficulty, if there is one, would appear to be purely verbal, and one which a little ingenuity should readily resolve.

Finally, there is this: in the ordinary course of normal waking life we constantly make, or could make, such remarks as 'I am conscious of a hard table, a horrid stench, a squeaky noise, a dull pain, etc., etc.' Translated into terms of the language I am using, these remarks are equivalent to saying that certain tactile, olfactory, auditory, etc., cognita are compresent in a certain field of consciousness with certain others, these others notably including some of what we would call bodily origin (endosmatic sensa), linked with yet others forming a continuous group and constituting the life history of a certain body known as Whately Carington, and with a great number of memory images, etc., themselves linked therewith—the whole aggregate forming a reasonably coherent system, to the more immediately accessible constituents of which, at least, I refer as 'me', or in equivalent terms. I think it is probably correct to say that every conscious state known to incarnate man, or knowable by him, is of this general type; in particular, that the field invariably includes *some* constituents of bodily origin, however faint, vague, and marginal. At least it would be difficult to refute anyone who asserted that this was so. It might, therefore, be tempting to contend that since, as alleged, these bodily cognita (as I may call them for short) are invariable constituents of every conscious state, they are necessary constituents, and that there can be no consciousness without them. From this it might be further argued that consciousness is entirely body-dependent, and that mind—or consciousness—cannot exist or occur except in conjunction with it.

[1] Except perhaps in the very far-fetched and highly Pickwickian sense that we might observe a deformation of the acoustic field due to the presence of a receptor-organ.

Any such contention would be, it seems clear to me, wholly fallacious, and I mention it only for immediate dismissal, on the simple ground that experience shows no kind of correlation between the intensity or degree of consciousness and the extent to which somatic cognita are present in the field. If anything, I should say, the contrary is the case. There are, of course, occasions, as when we have injured ourselves or are ill, when almost the whole field is taken up with our aches and pains—i.e. we are conscious of little else than our bodies—but equally, or perhaps more so, there are others, when we are intensely preoccupied with external scenes, sounds, or activities, or with the most abstract kind of thinking, when somatic cognita fade right away to the extreme margins of the field or out of it altogether; and I think most people would agree that, of the two, the latter type of state is that in which we are the more acutely conscious, though not *self*-conscious.

This brings me to one of the most important points, in its own way, that I wish to discuss, and one to which I must devote a separate section.

72. *Self-consciousness.* In the course of his extremely kindly review of my book *Telepathy*, in which this view of the mind and consciousness was first publicly advanced, Professor Price complained that 'self-consciousness (a hard fact if ever there was one) seems to me to defeat it altogether.'[1]

To me, on the other hand, the answer seems so obvious that it can only have been a kind of linguistic double-cross that has prevented anyone else seeing it. Because, for once, the habitual verbal forms do correspond to the structure of the actual situation; but it so happens that the propositions they express thereby automatically become untrue. What we call 'self-consciousness' is indeed the hardest of 'hard' (i.e. indisputable) facts, and the verbal forms indicative of some kind of duality, in which something is related to something else ('I am conscious of myself' or equivalent) are perfectly correct; there is a duality in which something referred to as 'I' is related in a particular way, namely, that of being 'conscious of', something else, to wit, the referent of the word

[1] *Horizon*, Vol. XII, No. 67, July 1945, p. 58.

'self' in this context. But this very fact of duality—I think I should prefer the word 'dichotomy' here—automatically ensures that what most people presumably think of as 'self-consciousness' never occurs at all. Incidentally, complete introspection is, for the same reason, impossible.

I suppose the common, uncritical view would be that there is a mass of thoughts, memories, feelings, sensations, etc., which we call the 'self', and that somewhere or other there is a sort of little angel that sits up aloft and surveys this 'self' with detached and impartial eye—*alias* the 'I', the 'ego', or other metaphysical unobservable. What actually happens is that the condition known as 'self-consciousness' arises *only* when we are in a more or less dissociated state, in which one part of our total self (or mind) is separated from or set over against another part, so that there is that kind of 'tension' or whatever you like to call it between them which, according to me, *is* 'consciousness' or the state of 'being conscious'. A completely integrated and non-dissociated mind can never be self-conscious.

This difference between consciousness and self-consciousness is, I believe, of the most extreme importance, not only —or even not so much—at the technical level, but in the most fundamental connexions; in such connexions, that is to say (though I shall not enlarge on them here), as those discussed under the headings of 'the object of living', 'the Purpose of Life', or 'the Spiritual Destiny of Man'. So far, of course, as any meaning at all can fairly be ascribed to these forms of words.

The really essential point is almost perfectly given in Mr. Hilaire Belloc's well-known lines about the water spider:

> He moves upon the water's face
> With ease, celerity and grace;
> But if he ever stopped to think
> Of how he did it, he would sink.

Self-consciousness, in fact, would spoil the performance, as it inevitably spoils every other, from playing the violin to living the good life. But it will be better to discuss the matter in more human and fractionally more formal terms.

Imagine yourself doing anything you can do really well, particularly in conditions which require intense concentration on the activity concerned. Perhaps you are a good car-driver, and you are handling an inferior car, with bad lights, worn tyres, and indifferent brakes—in a hurry—over a greasy and unknown road in the dark. Just the kind of situation you are good enough rather to enjoy than otherwise. Intensely concentrated, you do not think *about* what you are doing—you do it. Every patch of light and shadow, every hump or pothole or variation in camber, is interpreted and reacted to, every incipient slither is automatically corrected; you nurse your engine, caress your steering, sensitively feel your brakes, and unthinkingly employ every dodge and artifice that your years of assorted experience have taught you. And the same applies to more highbrow activities, from the performance of a Beethoven sonata to the dissection of a spider's brain. The point is that, in all such conditions, you are eminently—perhaps pre-eminently—*conscious*, but not in the least *self*-conscious. If you are, things go wrong. The moment he begins to think about himself, the driver will probably start messing his gear-changes, getting heavy on his brakes, and the car may well end up in the ditch; the anatomist will ham-handedly ruin his work, the concert performer will break down and rush weeping from the platform. And, broadly speaking, the moment we being to think about what sort of impression we are making, we begin to make a bad one.

In the above passage I have been discussing particular and limited activities, but the same principle applies on the widest scale. In these specific instances of car-driving, piano-playing, etc., there is, if I may put it so, perfect integration of a single limited system. All your relevant experience of the activity concerned (and perhaps, according to me, experience derived telepathically from other minds also—incarnate or discarnate) conspires, so to say, to determine your behaviour, which is the nett resultant of it. This mass of experience (cognita) is associated into a perfectly integrated system in which there is no conflict and no dissociation, but in which every item plays its balanced part in determining your behaviour; and that is why, within the limits set by the extent of your experience, your performance is perfect.

But the same applies to the living of life as a whole. In practice, one's mind (at least I speak for myself and probably for many others) is a lamentably disunited aggregate of non-integrated systems, some of which pull one way and others another, with the result that we are always 'checking and balancing' (like the American Constitution), wondering and deciding, thinking about what Mrs. Jones will say, or whether such and such a course will 'look well'. If only we could integrate all these systems with each other, so that every action was the unself-conscious resultant of the whole of our experience, not weighed or judged but just working, we could at least go through life in unconflicted self-harmony—though probably our fellow men would promptly lock us up as being too simple-minded for this world.

73. *Self-consciousness, contd.* I do not know whether I have made myself at all clear, and the point is one which it is difficult to emphasize except by bare reiteration. Professor Price, and doubtless most other philosophers also, wants 'an ego or *Atman* (spirit, self) which owns' the psychons (cognitum groups) 'and is aware of them'. I don't mind; let's have an ego or anything you like. But let it be so defined, and let statements about it (including the assertion of its existence) be such that evidence as to their truth or falsity can be adduced. It is, I submit, no use saying that there 'must' be an ego, because otherwise there could not be self-consciousness. This argument would hold only if it could be shown that, when what we call self-consciousness occurs, there is consciousness by something (the ego) of the *whole* self (though how this could exclude the ego I do not see). But all experience goes to show that nothing of the kind is true. It seems quite clear that dissociative processes can easily account, in principle, for self-consciousness of some kind and in some degree, whereas there is nothing to suggest that this self-consciousness is ever complete. On the contrary, the part of the total 'self' which the remainder may be said to be conscious 'of' seems nearly always to be only a small fraction

of that total. It may be no more than a yellow patch, as in my imaginary interview with the lemon (for this, as part of my contemporary field of consciousness, is part of my mind and so of my 'self'); or it may be those particular cognitum groups and sequences, present or prospective, constituting my behaviour on the lecture platform, which are at variance with those constituting the idea of the behaviour I should like to display, and give rise by this discordance to the feelings of shyness, etc., which embarrass me. But these, or those others involved in any other case, are but small samples from the total cognitum system which makes up my mind or self. In the relatively rare instances where the schism, so to call it, divides the mind or self into something like equal parts, we have the curious and apparently very unpleasant condition of the victim not being sure which of them is his 'true' self, or whether either is. The case already mentioned of the Rev. T. C. Hanna[1] affords a good example of such a struggle between conflicting selves, and I understand that a somewhat similar experience was reported by Swedenborg, though I have not read this.[2]

I see no reason whatever for invoking anything beyond dissociative processes to account for the phenomena of self-consciousness, whereas all the previous remarks about unobservables apply equally to this aspect of the ego. If you cannot observe it, it does not exist; if you can, then it is some kind of a cognitum pattern.[3] Moreover, if you cannot define

[1] Sidis and Goodhart, *Multiple Personality* (London, 1905). See *Proceedings*, of the Society for Psychical Research, Vol. XIX, (1905–7), pp. 345 ff. and Vol. XXVI (1912–13), pp. 266, 284.

[2] Cf. S. Toksvig, *Emanuel Swedenborg* (Yale University Press, 1948), pp. 180–1. (H. H. P.)

[3] Perhaps I ought to guard yet once more against the contention that one can 'infer' the existence of objects, etc., which one cannot observe, e.g. an invisible planet from the perturbations it produces in others. But all this means is that you say that *if* a sufficiently powerful telescope be turned in the right direction at the right time, you will cognize the appropriate visual cognita. If, under these conditions, the visual cognita are not obtained, then you must either make up an *ad hoc* story to account for this (e.g. that the planet is black) or abandon the hypothesis, as in the case of the hypothetical planet Vulcan once postulated to explain the aberrations of Mercury, and say frankly that the extra planet does not exist and that the effects are due to some other cause (relativity theory, etc.).

it, i.e. state its properties, it is no use talking about it; and if you can, then you must do so in terms of cognitum sequences, etc., of some kind. The ego, I insist, like the *Ding*, is a purely metaphysical entity gratuitously introduced to cover up the fact that it is sometimes difficult accurately to state certain phenomena in terms of cognitum sequences alone.

If the reader bears in mind that there is a difference between being conscious and being self-conscious, and also what I have suggested about dissociation as a requirement for the latter condition, I think his own experience will soon convince him that the facts can adequately be covered by what I have said, without dragging any metaphysical unobservables into the story.

There are, of course, many points which it would be of some technical interest to discuss here, notably whether the sub-systems of the mind—from sentiments and complexes up to full-fledged secondary personalities—can properly be said to be conscious in their own right, or to have, as we should say, any conscious life of their own. I think that, with due caution, there can be little doubt that they can and have. But, as I have already said, 'consciousness', or 'being conscious' is a matter of degree and depends, among other factors at least, on the extent and complexity of the psychon system concerned—much as 'being alive' is a matter of degree (in the sense that a slug is more alive than an amoeba, and a man than a slug). I cannot suppose that a relatively simple psychon group such as a psycho-analytic complex is in any high degree conscious, though it may be very active and its effects far-reaching (as is true of certain lowly forms of life), but one can hardly deny consciousness to so relatively well developed a system as, say, a mediumistic control or the 'Sally' who plagued the normal Miss Beauchamp.[1] But these are purely technical matters having nothing to do with the main discussion on which we are engaged. Note, by the way, that this applies to almost the whole of this chapter. Once we have accepted the manifest fact that the word 'mind' must

[1] Cf. Morton Prince, *The Dissociation of a Personality* (Longmans, Green). See also App. II, below, p. 239.

either refer to cognitum groups, sequences, etc., or some kind, or be meaningless, the rest is only a question of working out as best we may, in the light of our very limited knowledge, what kind of regularities (laws of psychology) do, in fact, characterize the phenomena (sequences) which we agree that it is convenient to call 'mental'.[1]

74. *Emotions and Will* (1). I always get into more or less serious trouble when I attempt to discuss Emotion, and I shall do no better here, for I must confess that I am entirely intransigent on the subject, and regard all alternative views (detail apart) as no more than themselves the product of emotional states induced by human insistence on thinking of 'emotion' as something magical and sacrosanct and not to be desecrated by the rude touch of intelligent thinking. And it will be even worse when I come to speak of the Will.

The progress of psychology has been perhaps in some respects factilitated, but on the whole very much retarded, by the arbitrary division of states of mind, or components thereof, into Cognitive, Affective (emotional), and Conative (willing or striving). This is worse than basing the science of Ethnology on the alleged descent of all men from Shem or Ham or Japhet; for these at least were all human beings, whereas the classification of psychological states under the three headings just mentioned has tended to suggest that these refer to conditions inherently different *sui generis* from each other.

Let me be dogmatic and say firmly that this is all rubbish. Properly speaking, all states are cognitive; the differences, which, of course, it is valuable to distinguish for certain purposes, are to be found solely in the character of the cognita cognized. To say that a certain state of mind is purely cognitive is (in the sense in which psychologists would say it) usually and perhaps always not quite accurate; but in so far as it approximates to correctness it is merely equivalent to saying that the field of consciousness concerned contains a

[1] [Those who can bear to read 'metaphysical' works will notice that Carington's theory of self-consciousness is curiously like F. H. Bradley's in *Appearance and Reality*, Ch. IX. Cf. especially Sect. 6 of that chapter. (H. H. P.)]

negligibly small proportion of cognita of that particular type which, when present in relative abundance, would characterize the state as emotional. And similarly for conation.

These characteristic cognita are sensa of, as we would ordinarily say, endosomatic origin, notably visceral in the case of emotion and probably intramuscular, etc., in the case of conation, or memory images of these. We all of us know the sinking feeling in the pit of the stomach, the dryness of the mouth, and the feebleness of the knees characteristic of what we call the emotion of fear; or the sensations of tightness and heat about the head which accompany anger; and most people when urged to 'will' something react by clenching the teeth, contracting the diaphragm and furrowing the brow. But it is incorrect to say that these sensa are caused by or accompany the emotion, etc. In so far as such a phrase is applicable at all, they *are* the emotion; or, more accurately, they are the distinguishing features of the states described as emotional in this or that respect.

At one time the commonly held view was that '. . . the mental perception of some fact excites the mental affection called the emotion, and that this latter state of mind gives rise to the bodily expression'. The James-Lange theory, on the contrary, which in substance seems to me the only sensible one, contends 'that *the bodily changes follow directly the perception of the exciting fact, and that our feeling of these same changes as they occur* IS *the emotion*'.[1] This is almost exactly what I have been saying, apart from minor differences of wording. James goes on: 'Common sense says: we lose our fortune, are sorry and weep; we meet a bear, are frightened and run; we are insulted by a rival, are angry and strike. The hypothesis here to be defended says that this order of sequence is incorrect, that the one mental state is not immediately induced by the other, that the bodily manifestations[2] must first be interposed between, and that the more rational statement is that we feel sorry because we cry,

[1] Quotations from William James, *The Principles of Psychology*, Vol. II, p. 449.

[2] Note that the 'bodily manifestations' must themselves be reducible to terms of cognitum sequences.

angry because we strike, frightened because we tremble, and not that we cry, strike or tremble, because we are sorry, angry or fearful, as the case may be.'[1]

Naturally, this sort of statement needs a good deal of modification before it can be accepted as satisfactory, and there has been a lot of quite unnecessary wrangling about it. Many writers have pointed out that the bare occurrence of a particular bodily manifestation is not enough, of itself, to produce or characterize a given emotional state. We may cry because we have a cinder in our eye or have smelled an onion, without being sad at all; we may strike without feeling angry; if we know that we can easily outstrip the bear, we may run in derision or exultation and not in fear. But this is only to say that the states of fear, anger, or sadness are characterized, not by some simple group of sensa (as from the lachrymatory glands, etc.), but by the adaptation of the body as a whole to the situation as a whole, to which the context of the specific stimulus may make all the difference. The working out of just what factors and relationships operate to make a state fearful or derisive, for example, is a matter for the physiological psychologists; it is of no special interest for our present purpose. All that concerns us here is that there is neither need nor justification for introducing metaphysical unobservables disguised as Essential Qualities of particular states or by any like terms. To speak of 'an emotion' being 'characterized *by*' such-and-such qualities is precisely parallel to postulating a hypothetical 'lemon-itself' characterized by the properties of yellowness, acidity, etc., which it is said to 'have'. Only cognita can be cognized, and if one state of mind is qualitively distinguishable from another the fact must be expressed, in principle, in terms of the differences between the cognita which compose them, or else the assertion of distinguishability is meaningless. The fact that it may be difficult to analyse different states or to identify their characteristic constituents has nothing to do with it.

75. *Emotions and Will* (2). Most people seem to dislike and resist the view of emotion sketched above, apparently

[1] *The Principles of Psychology,* Vol. II, pp. 449-50.

because they consider some emotions to be 'beautiful' or 'holy', etc., and feel that it is somehow derogatory to reduce, say, the emotion of pity too literally to a matter of 'bowels', even of compassion. This is, of course, a very common attitude, but one which cannot be too strongly repudiated. It adds nothing to the beauties (if you consider them so) of a sunset to suppose that the colours are due to blushes on the cheek of retiring Phoebus; it in no way detracts from them to be able to describe what is happening in terms of the differential scattering of light by atmospheric particles, etc. The colours are what they are regardless of how they come about, and the same applies to human emotions and human behaviour.

But strong as this resistance often is in the case of emotion, it becomes positively fanatical, not to say ferocious, if the Freedom of the Will is called in question, and defence of this conception has ranged from Dr. Johnson's dogmatic 'We *know* our will is free, and that's an end on't' to the massive tomes of innumerable metaphysicians.

I myself unfortunately, have never been able to ascertain what the phrase Free Will means; and I do not believe that anyone else has either. The doctrine of Determinism I can understand well enough; and the fact that it appears to be fundamentally false does not prevent my doing so. With the notion of Chance I am at least familiar, though I doubt whether it has received as close a study as it deserves. But such statements as 'My Will is Free', 'We have Freedom of Will', etc., appear to me to mean exactly nothing at all. I conclude that all discussion of Free Will has been nothing but a logorrhagic extraversation of meaningless non-sense.

Consider: the classical conception of determinism, originated, I think, by Laplace, was at any rate simple enough. The universe is subject, it was contended, to Law; its whole configuration at any moment, from the celestial bodies to the molecules in my brain, is the inevitable consequence of its configuration at the immediately preceding moment. Thus a miraculously endowed calculator, knowing the configuration at any one moment, the positions, velocities, etc., of every particle, and the laws which govern their movements, would be able to predict with complete certainty what the exact configuration would be at any other moment, however remote—down to the fact that Whately Carington, at 10.51 a.m., G.M.T., in the latitude and longitude of Sennen Cove, on this fourth day of December 1946, would scratch his right ear (as I have just done) for no apparent reason.

If this contention were correct, then all our actions would be completely predetermined, so that it would be idle to talk about choosing one course of action rather than another, or of deciding to do good rather than evil, or vice versa, and the whole basis of human morality (which is responsibility) would vanish away. The conviction that this was true weighed, as Dr. Stebbing says, like a nightmare on the great Victorians of the nineteenth century,[1] and many attempts of one kind and another were made to evade or refute it. But the more deeply scientists delved into the mysteries of Nature, the more all-pervading and ineluctable did the reign of law appear to be.

This state of affairs endured, and indeed worsened, for some two hundred years, till the introduction of Heisenberg's famous Principle of Indeterminacy in 1927 brought the whole edifice of strict physical determinism crashing for ever to the ground. Laplace and his followers, like all logicians who do not actually make errors in reasoning, was, of course, perfectly right—given his premises. IF you knew the exact position and momentum of every particle (electron, proton, etc.) in the universe at any moment, you *could*, in principle, calculate their positions, etc. (i.e. the exact configuration of the universe) at any other moment; and the fact that any such procedure was in fact quite impracticable had always been rightly deemed irrelevant to the logical issue. But Heisenberg showed that it is not merely impracticable, but *inherently impossible*, to determine both the position and momentum of any particle with exactitude,[2] and that any increase in

[1] The interested reader should consult her book, *Philosophy and the Physicists*, already mentioned, for an excellent outline of the subject.

[2] So that, incidentally, it is meaningless to say that an electron, for example, *has* both exact position and exact momentum.

accuracy of determining the one involves a corresponding decrease in accuracy in determining the other. And there is no question as to his being right in this, though I do not propose to go into technical details here. Thus the all-important premise on which the deterministic argument rested was destroyed, and the doctrine of strict physical determinism with it. It is no use arguing that it is possible to deduce a future configuration of the universe from the present configuration, if you cannot ascertain what the present configuration is; and any uncertainty, however small, will ruin your argument, just as the lady's reputation was none the less impaired by the fact that her baby was 'only a very little one'. Heisenberg's principle, and the substitution in general of statistical laws for the classical formulations in the light of quantum theory, make no worthwhile difference to large-scale phenomena; but when we are dealing with atomic or sub-atomic occurrences, such as (it is commonly supposed) might affect a chemical change in a brain-cell and thus a whole course of overt action, they may be of the highest importance.

These facts and considerations led some enthusiasts to hail the Principle as exorcizing the nightmare and opening the door at least by a useful crack to the reintroduction of Free Will. Actually, as we shall see, it does nothing of the kind; for the antithesis is not of Determinism to Free Will, as is commonly but erroneously supposed, but of Determinism to Chance—with the reservation that I am not at all sure whether 'Chance' is a genuine alternative to Determinism.

76. *Will, contd.* The really extraordinary thing is that no one, to the best of my knowledge and belief, has ever stopped for a moment to ask what all the discussion is about; that is to say, to ask what the phrase 'Free Will' *means*.[1] Everyone seems to take it for granted that there can be no doubt about this; but the assumption appears to me to be wholly without justification.

Consider again: I think there is no doubt at all about what

[1] [Some philosophers certainly have. Cf. for example, Locke's *Essay Concerning Human Understanding*, Bk. II, Ch. XXI, Sects. 14 ff. (H.H.P.)]

we mean by acting under compulsion. If we express too frankly views which happen to be unpopular with our neighbours, there is a serious risk that they may 'take us and carry us whither we would not'—or force us to march there under pain of death. And we are presumably all of us only too familiar with the experience of, as we say, 'wanting' to do one thing, but being 'obliged' to do another. In the one case the compulsion is external, in the other we should describe it as originating from within. In the latter circumstances we can, as a rule, quite clearly distinguish two conflicting or incompatible systems of tendencies, desires, etc., of which one overrides the other. I do not see how anything that can fairly be called 'Will' enters into either of these situations.

Even in cases where we would normally be said to make 'an effort of will', I still cannot find the referent of the word 'Will'. I lay in bed this morning, feeling intensely desirous of turning over and going to sleep again; at the same time I was also desirous of getting on with the job of writing this book. Finally, I made one of these famous efforts of will, and forced my reluctant body to crawl from between the sheets. But surely all that was happening was that I was aware of the conflict between two mutally exclusive 'pressure groups', as one might call them, of incompatible desires, one of which finally prevailed.

More generally, whenever an event occurs which we describe as the result of deciding, choosing, selecting, willing, etc., our action seems always to be taken for some *reason*, or nett balance of reasons, and the word 'reason' is only a shorthand name for a psychon-sub-system or pressure group of one kind or another. The reasons may be good or bad, important or trivial, numerous or few, but there seems to be always *some* sort of consideration which turns the scale one way or the other. In so far as this is correct, then we are subject to a psychological determinism just as effective as the older physical determinism; and if anyone can show that there are cases in which there is actually *no* determining factor of any kind (which I do not think I believe) then one would have to fall back on saying that the action concerned is a matter purely of Chance—though this would involve difficulties of its own, since it is difficult to define 'chance' except in terms of a large number of randomly acting 'causes', or on like lines.

And if it is impossible to find any referent of the word 'Will', it is even more so in the case of 'Free Will'—and, incidentally, what on earth could be meant by '*un*free will'?

Exactly the same considerations would apply even if we were to allow ourselves the illegitimate luxury of postulating (or ostensibly inferring) the existence of an Ego, endowed with the special property of choosing, deciding, willing, etc. Either this Highest Authority would act rationally, i.e. for some reason or another, in which case it would be itself determined and we should only have pushed our difficulty one stage farther back; or else it would not, in which case it would be no more than a kind of psychic roulette-wheel adding nothing that I can see to our conception of human dignity.

As so often, it is more profitable to consider why people hold the views (or at least utter the noises) they do than to inquire into the logic of their alleged opinions, and the present muddle seems to be due to a number of factors, all of which I think are manifestations of vanity and individualism. For example, it is generally agreed that if we deny 'free will' we have to forgo the right to condemn and punish other people (in any retributive sense at least); and since condemning other people makes us feel superior ourselves, this would be a serious privation to most of us. Again, nearly everyone likes to feel 'I am the master of my fate, I am the captain of my soul' and resents the suggestion that he is merely a determinate mechanism. (But why 'merely'? It may be a very good thing to be.) But the most important factor of all, I suppose, was the identification of 'determinism' with *physical* determinism—which is not at all the same thing. Physical determinism, with its ancillary doctrine of the abrupt cessation of consciousness at the death of the body, outraged and antagonized that urge to live which is the most fundamental instinct (except possibly the urge to reproduce) of all animate creation. But if 'Free Will' could be maintained, with its implication that Mind or Consciousness could, on occasion at least, dictate the course of physical events, then Matter would not have the last word, and material dissolution need not necessarily be the end.

Thus the resistance to physical determinism, and the desire to support Free Will at all costs (and to blazes with quibbles about meaning), was essentially directed against the physical rather than the deterministic aspects of the doctrine; and those concerned failed to see that, even if 'mind' may directly or indirectly influence the course of physical (material) events, as I have little doubt that it may, this fact in no way solves the difficulty (if it is one) of determinism. We do not get rid of determinism merely by pushing the problem back from the field of matter into the field of mind.

No. If we are to talk sense about this business at all (and it is admittedly difficult to do so) we must adopt a quite different line of approach. To speak of 'the Will' as if the word referred to some kind of active agent or operative force is meaningless, since no referent can be found for it.[1] But, on the other hand, there are plenty of clearly identifiable states commonly described as states of willing, striving, choosing, deciding, etc., and it is worth while examining what is actually going on in these conditions, or in such of them as interest us from the point of view of 'will' and particularly of 'free will'.

The trouble here is that all these words, and especially the word 'conative' already mentioned, seem to be loosely and vaguely used to refer to states involving, in varying proportions, at least two clearly distinguishable elements. We may speak of these as 'effort' or 'strain' on the one hand, and 'decision' or 'choice' on the other. Consider the cases of a man (*a*) striving to lift a heavy weight, (*b*) attempting to solve a mechanical puzzle, such as picking a lock, (*c*) struggling against temptation, (*d*) trying to make up his mind whether to sell some shares. All these would probably be spoken of as 'conative' states, and all could be expressed in terms of

[1] ['The question is not proper, whether the will be free, but whether a man be free', Locke, *Essay*, Bk. II, Ch. XXI, Sect. 21. (H. H. P.)]

'trying'; but they are obviously markedly dissimilar, and the conception of 'free will' would hardly be applicable to any but the two last. In the first, the element of strain or effort is predominant, and that of choice or decision is negligible, except in so far as the man has the option of discontinuing his activity. In the second, there is little or nothing in the way of external resistance to be overcome, or opportunity for the exercise of 'will' or 'free will' in any ordinary sense. In the third, decision and 'will' would commonly be said to be involved in a high degree, and there will be, *ex hypothesi*, a more or less intense conflict between opposing psychon systems or pressure groups constituting the desires to do or not to do whatever may be concerned. And in the fourth, there will again be conflict between pressure groups of ideas (psychons), though the word 'will' would not ordinarily be applied in this connexion. We can, of course, say that opposing pressure groups can be identified in the first two cases also, in the form of sensations of discomfort, etc., tending to make the weight-lifter let go, or of boredom, mental fatigue, or the like tending to make the lock-picker abandon his task, or, in this latter case, differing sets of images, etc. representing alternative procedures that he might adopt.[1]

But I do not think we need go into elaborate analyses here. The point I want to make is that those experiences characteristically describable as the exercise of 'free will' invariably involve not only, or necessarily, the element of effort or strain, but one of 'selection' or 'choice' between two or more possibilities; that is to say, they invariably involve the conflict or opposition of two or more incompatible or mutually exclusive psychon-systems or pressure groups.

If there is no such conflict, the question of Free Will simply *does not arise*. In so far as such an assertion as 'Our

[1] It is not quite so easy in the case of someone 'willing', for example, that a die shall fall with one face rather than another upwards, or a would-be magician willing that an object shall rise in the air by other than mechanical means. But I suppose that even here there may be considerable opposition from systems representing the conviction that it can't be done.

Will is Free' purports to say anything beyond denying (almost certainly correctly) the narrow doctrine of strict physical determinism, and thereby shifting determinism from the material to the mental realm, it is meaningless non-sense.

I think I would go further and categorically challenge any exponent of Free Will, or indeed any opponent of that conception either, to say what he means when he affirms (or denies) that this is a Free Will Universe, or makes any statement to like effect. What test does he propose to apply in order to verify or disprove any such statement? What process of observation can be prescribed whereby a Free-Will Universe can be distinguished from a non-Free-Will Universe? After all, any such statement must purport to be concerned with a matter of fact capable of translation into terms of empirical observation; and if none can be made which are relevant to its truth or falsity, then it is meaningless. I submit that the only way in which such a challenge could be met would be by claiming that we do in fact, and as a matter of common experience, 'exert our Will' or 'exercise our power of choice'. But this is only equivalent, as I have tried to indicate, to saying that we enjoy (if that be the proper word for it) certain experiences in which we are aware of two or more conflicting pressure groups of which one finally outweighs the others. The word 'Will', in so far as it refers to a matter of choosing as distinct from mere sensation of effort, is only the name we give to our awareness of the process and outcome of such conflict.

Although I should describe myself as a hundred per cent determinist (with possible reservations about 'chance' which make no essential difference that I can discern at present) and fail to see what other conclusion one could rationally come to, I do not feel that this view gives any occasion for alarm or despondency, unless we happen to entertain an unnatural passion for irrational and chaotic behaviour.

The whole position seems to me to be very much on all fours with Prince Kropotkin's conception of the ideal community, in which everyone does exactly what he likes, yet everything proceeds in a perfectly orderly fashion, because everyone is so integrated in that community with everyone else that he never even wishes to do anything inimical to its best interests; or, alternatively, everyone so fully understands what is going on, and the consequences of potential actions, that it never occurs to him that there is more than one way of behaving. Similarly, in a mind perfectly integrated within itself and harmonized with others, there could be no conflicts and, therefore, no experiences of willing and choosing. Laws are only irksome in so far as we wish to break them; willing and choosing are but symptoms of kicking against the pricks; and in the limit there is no distinguishable difference between doing what you like and liking what you do.

77. *Limits of the Mind: Individuality.* We have pictured the mind as a system or aggregate of cognita or cognitum groups (conveniently, for some purposes, called 'psychons') organized as regards their interrelations and sequences, etc., according to the laws (observable regularities) of psychology, or at any rate according to laws other than those of material physics. The question now arises of what it is that sets limits to what we call the individual mind; and the short answer to this is that there is nothing, and that there are no limits. More correctly, perhaps, the limitations of the individual mind are limitations *de facto* and not *de jure.*

Earlier in this chapter I denounced as metaphysical and meaningless (because inherently unobservable) the notion of a mind-itself acting as a kind of 'container' for the ideas, thoughts, feelings, or cognitum groups generally, which would colloquially be said to be 'in' it. I used this crude word 'container' more or less deliberately and because I wanted it to sound as nonsensical as it is. But exactly similar considerations of inherent unobservability will apply to any other more sophisticated conception of some entity or mechanism postulated *ad hoc* to explain the cutting off of one mind from another. As always, any such 'cutting off' is a phenomenon which must be stated in terms of cognitum groups and sequences, or be meaningless.

I have also been deliberately vague as regards my use of the terms 'mind' (in general) and 'the mind' (meaning a typical individual mind). This is because we are not entitled to draw any sharp fundamental distinction between 'mind in general' and 'the individual mind', *unless* we cheat by postulating some metaphysical and unobservable entity or mechanism to make the distinction valid.[1]

What we call the individual mind is, in fact, far from indivisible; it is an aggregate of psychons, forming a relatively large system, but consisting of numerous sub-systems—moods, sentiments, complexes, etc., up to (sometimes) full multiple personalities, of varying size, stability, and coherence, federated into what we call 'So-and-so's mind'. In the same way, what we call 'mind in general' is not, as the term is too apt to suggest, some kind of a uniform and homogeneous substratum or all-pervading constituent (like the old-fashioned 'ether') of all mental activity. It is rather to be thought of as a super-aggregate of *all* cognita, organized or patterned otherwise than according to the laws of material physics, within which certain relatively stable condensations, so to call them, known as individual minds, are observable (or inferable), and within these again sub-condensations or systems (the sentiments, complexes, etc.) are to be found.[2] To use a very rough analogy, we might say that 'mind in general', or 'Mind', if you prefer this, corresponds to all the astronomical bodies collectively, the individual mind to a particular galaxy, star-cluster, or spiral nebula, and the complex, sentiment, etc., to a particular star and its planets, if any. We do not say that there is any kind of a 'container' or magical 'ring-pass-not' which circumscribes the galaxies and prevents them mingling with each other; we say that the gravitational forces (or curvatures of space) are such that they do, in fact, aggregate

[1] Cf. the author's *Telepathy*, pp. 118–20. (H. H. P.)

[2] For aught I know, there may be cognita or cognizables that are not organized according to any regularities (laws) at all, but I do not think we need worry about this possibility here. Also, there may be—and I should think quite possibly are—super-individual minds intermediate between what we call an individual mind and the total aggregate; but this suggestion is liable to lead to unwarranted and unverifiable speculation unless carefully controlled. I shall have something to say about this when I come to discuss Group Minds below (cf. Sect. 82, pp. 214 ff.). Cf. the author's *Telepathy*, Ch. XII.

and cohere in the manner observed and that particular stars do not, in fact, migrate from one galaxy to another.[1] Similarly, we are only entitled to say that whatever degree of separateness or cutting off is, in fact, observed between individual minds is due, not to the presence of some magical containing entity, but to the operation of those 'forces' of association and dissociation which do, in fact, operate between psychons and psychon systems—just as we speak of these forces operating to separate complexes or secondary personalities with an individual mind.

The foregoing is not, I think, quite so much of a dogmatic pronouncement by myself as might appear at first sight. Prejudice apart, it all seems to me to follow naturally from what I have said, which is all that it seems permissible to say with assurance, about the nature of those non-material phenomena which we commonly agree to call 'mental' or 'of the mind'. Once it is agreed (and I fail to see how it can be disputed) that all 'mental events', 'states of mind', 'contents of consciousness', etc., must be reducible to terms of cognitum groups and sequences, and 'forces' acting between them,[2] then some such view as that roughly sketched above of individual minds as themselves aggregated in a kind of superaggregate, with the same forces acting between them as within them, seems to me virtually inevitable. Any alternative view of individual minds as being—to use Dr. Gardner Murphy's graphic term—'encapsulated' entities, *necessarily* isolated from each other, would involve *defining* a cognitum group or aggregate, etc., in such a way that any statement implying non-encapsulation or non-isolation would constitute a contradiction in terms; and I do not at all see how this could be done, except by an *a priori* postulation of a metaphysical

[1] I doubt whether it would be strictly correct to say that such migration is flatly impossible, though of course the probability of its occurrence would be untra-unimaginably infinitesimal.

[2] Remember carefully what I said above about the illegitimacy of thinking of 'forces' as existents independent of the entities on which they are said to act or operate. A 'force' *is* (more accurately, the word 'force' is a name for) the movements, changes, etc., in the relevant observables, inasmuch as anything said about it must be reducible to statements about observations made on them. (Cf. Sect. 67, p. 165, above.)

unobservable of some kind; and this would be illegitimate for the reasons I have so often emphasized.

78. *Interaction of Individual Minds.* What I am trying to do, though I do not know with what success, is, as always, to get rid of preconceptions arising out of verbal habits operating on insufficient factual knowledge. The whole business is one of fact, not of gratuitous metaphysical apriorism. Even if, following the metaphysicians, we contrive to cook up some sort of a definition which, so to say, ensures (or appears to ensure) the rigid isolation of individual minds, we must still (unless we wish to fall into the same pits as they) take steps to ascertain whether entities conforming to our definition do, in fact, occur in nature. But this is rather by the way, and the operative word here is 'gratuitous'.

We all know, as a matter of immediate experience, that psychons (cognitum groups) do act on each other within the so-called individual mind (as in instances of association), and this fact of acting on each other must be taken as an observable mode of behaviour or 'property' of psychons. It is quite uncalled for to assume that two psychons have the property of acting on each other when they are both, as we say, 'in' a certain mind, M_1, but not when one of them is in M_1 and the other in M_2. Unless we go out of our way to invent some reason why they should not, which is equivalent to introducing some sort of hypothetical and metaphysical entity so long as we are talking *a priori*, it seems at least natural to assume that psychons said to be in different minds may interact as freely, in principle, as those said to be in the same mind.[1]

But what do we mean by being 'in a mind'; and can a psychon be 'in', or 'form a constituent of' two minds at once, and if not, why not?

[1] Note that I say 'in principle'. Every particle of matter in the universe attracts every other particle, but not in the same degree. The attraction varies with the masses of the particles and with the distance between them. If we find that there are quantities corresponding to mass and distance in the case of psychons we may conclude that psychons in different minds interact more feebly than psychons in the same mind; but that will not affect the principle that they can interact.

If we reject the notion of an unobservable 'container', or equivalent mechanism, it seems to me clear that all we can say about these questions can and must be expressed in terms of probability.[1] A cognitum group, or psychon, can (and must) be said to be 'in' or 'be a constituent of' or 'belong to' a mind if there is a finite probability of it appearing in or forming a part of the relevant field of consciousness.

Note that I say 'relevant field', and that this is properly the starting-point rather than the 'mind', which is, as it were, the source from which the psychons in the field at any moment are drawn. The mind belongs to the field of consciousness rather than the field of consciousness to the mind. In normal waking life our field is that of which bodily cognitum groups (sensa of exo- and particularly of endosomatic origin) form a semi-permanent nucleus (the 'vague mass of bodily feeling', etc.). But, as already indicated, I see no reason to doubt that other condensations or systems have, in their degree, fields of consciousness of their own.

Now, unless we go out of our way to ensure it by means of some kind of arbitrary definition *a priori*, we cannot legitimately say that the probability of any given cognitum group appearing in a particular field of consciousness is actually zero—i.e. that its appearance is flatly impossible—though the probability may be very small and the appearance very unlikely. To speak rather colloquially, but I think quite unambiguously, it is a question of how closely the cognitum in question is associatively linked (directly or indirectly) with the cognita actually forming the field; and, in the absence of artificial barriers erected by definition, it is impossible to say that there is *no* linkage. I conclude, therefore, that any cognitum, cognitum group, or psychon whatsoever, if all that exist, *may* in principle form part of any field of consciousness.

That is to say, there are, in principle, *no* limits to what we commonly call the individual mind, though in practice, at our present stage of development, it is to all intents and purposes severely circumscribed. To speak of my mind or yours is,

[1] Cf. Sect. 68, above, pp. 167–8.

strictly, to speak of a single class of entities, to wit all existing cognita; but for practical purposes the possessive adjectives 'my' and 'yours' restrict the referent to those cognita which are vaguely but sufficiently describable as 'reasonably likely' to appear in our respective fields of consciousness, though I think we should importantly include those which in appreciable degree determine whatever does appear.

79. *Interaction of Minds, contd.* (2). We must now consider the question whether the same cognitum, cognitum group, or psychon can form part of two or more minds (as roughly circumscribed above) or enter into two fields of consciousness simultaneously.

The plain and obvious answer is that there is no conceivable reason why it should not, once we have got rid of the superstition that minds must be localized in different places with barriers between them. To say that a given cognitum is part of my mind is merely to say that it is more or less closely linked with certain other cognita in a more or less coherent system, such that, when one or more members of that system become themselves compresent with certain others (particularly my 'bodily' cognita) in what is commonly called my field of consciousness, any other member of the system (including, of course, the 'given' cognitum) is more likely to be also compresent than are other cognita less closely linked with that system.[1] And there is nothing whatever to be said *a priori* against a cognitum being linked with as many others as we please, some of which may themselves be linked into my system, others into yours, or all of them into different systems. Unless, again, we gratuitously set up artificial restrictions to exclude this from possibility.

Moreover, in certain circumstances at least, we may invoke the principle which has often served us before, and point out that it is illegitimate and meaningless to declare two entities to be different unless their difference can be demonstrated, or evidence relevant to the assertion adduced. Entities which

[1] This statement is, of course, circular because closeness of linkage can only be measured or graded by the observed relative frequency (probability) of compresence or close following. But that is irrelevant here.

are inherently indistinguishable in practice are not to be called 'different' unless two or more of them are simultaneously observable.

Suppose you and I are walking on a cliff and both hear the cry of a gull. Assuming that we are equally acute of hearing,[1] that the intervening atmosphere is homogeneous, that we are equidistant from the source of sound, etc., can any meaning be attached to the statement that we cognize different auditory cognita? In the ordinary way, and on the further assumption that our descriptions of the sounds (carried to the ostensive level, if need be) are identically similar, I should say that no meaning could be attached to the statement, on the ground that it seems inherently impossible to test it empirically. But just at this point, when we are getting nicely out of our depth and beginning (I speak for myself at least) to flounder in a sea of speculation, the possibility of deduction and experiment comes to our aid.

Let us further suppose that at the moment when we both hear the cry of the gull I happen to see an adder cross the path, but that you, looking in the opposite direction, do not. I say nothing about it for fear of alarming you, and in due course we go home and have tea. Now the facts of association lead us to expect that, next time I hear the cry of a gull, I am more likely to think of an adder (as we put it) than if I had not enjoyed the two experiences in conjunction; but, in the ordinary way, we should not expect *you* to be more likely than before to think of an adder next time you hear the cry of a gull. If you do (or if I do *mutatis mutandis* in comparable conditions) we would say that there was something odd about it, e.g. that it was a queer coincidence. But if this kind of thing happened so often that coincidence could, to all intents and purposes, be ruled out, we should say that (using ordinary language) there must be some causal connexion between

[1] Considerable complications naturally arise if we are not; notably the question whether, to put it rather roughly, two sounds of differing intensity but otherwise identically similar are to be regarded as radically different cognita, or as two groups resolvable into elements of intensity which are distinguishable and others (pitch, quality, etc.) which are not. But I do not think we need bother about this here.

associative connexions in my mind and certain sorts of mental events in yours.

Now consider the two alternative hypotheses, namely, first, that the gull-cry cognita cognized by both of us in the initial experience are different; second, that they are the same (or contain certain identical elements), i.e. that there is only one. On the first hypothesis my auditory gull-cry cognitum is associated with a visual adder-cognitum, while yours is not. In the light of the known facts of association we accordingly expect to find a tendency for visual adder-cognita to accompany or closely follow similar or identical[1] gull-cry cognita, on subsequent occasions, in my case but not in yours. On the second hypothesis there is only one gull-cry cognitum (or some element thereof) which is common to both of us and is, in fact, linked with an adder-cognitum, and we should expect to find a tendency for the latter to follow the former in all cases, yours as well as mine. This is not to say, of course, that the tendencies will be equal, i.e. that there will be an equal probability that adder will accompany gull-cry on subsequent occasions in the two cases. Any given cognitum is not associated with a single other, but with a large number woven into a highly complex system in which associative and dissociative forces play their part. My own tendency to think of adders when I hear gull cries will be only a tendency and by no means an invariable regularity, because innumerable alternative cognita will be associated with both; and in your case the gull-cry will also be associated with innumerable alternative cognita very many of which will be quite dissimilar to mine.

Still, on the second hypothesis, that there is only one cognitum (or element of a cognitum group) cognized by both parties, some such tendency should be observable if we look carefully enough; and, if we find it, the observation will constitute evidence in favour of this hypothesis rather than the other.

But, as a matter of fact, we do find quite fairly numerous

[1] I say 'identical' advisedly, because I see no means of empirically substantiating the contention that a gull-cry cognitum I cognize on one occasion is different from the identically similar gull-cry cognitum I cognize on another occasion.

instances of this general kind of thing happening—not usually of just the type described, but cases in which one person thinks, or dreams or has some kind of impression or vision of some event that is occurring to another or occupying his thoughts. When this happens to an extent that cannot plausibly be attributed to coincidence, we speak of telepathy, of which I shall have more to say later, and begin to speculate as to how it is that an idea 'in' A's mind manages to get 'transmitted' to B's mind. The systematic collation and study of spontaneous cases of such phenomena, over the last sixty years or so, backed more recently by the experimental work of various investigators, has now quite definitely established the facts as reliable;[1] and probably the only serious bar to their universal recognition is the lack of a generally acceptable theory to account for them.

I myself propounded the Association Theory in the work just mentioned,[2] and the foregoing passage is essentially an approach to the same conclusion by a different route. In utmost brevity the theory is this: when a certain idea (cognitum group) is telepathically transmitted, as we say from the mind of A to the mind of B, what happens is that this cognitum group or 'idea', X, is associated with some other cognitum group, K, in A's mind, and that if the idea K is presented to B (i.e. is present in his field of consciousness) it tends to call up X in the same way and for the same reasons (though normally in less degree) that it would if it were re-presented to A.[3] As explained in the aforementioned work, this theory seems to fit the facts well, and I see no reason for modifying it appreciably. But, although the point is trivial as regards the practical study of telepathic phenomena, there remained a certain ambiguity as to whether the K-idea

[1] See my book, *Telepathy* (Methuen, London, 1945).

[2] See also *Proc.*, S.P.R. Vol. XLVII, Part 168, July 1944.

[3] I use the term 'K-idea' to refer to any cognitum group performing this sort of function. In experimental work, I have suggested, it is the 'idea of the experiment' which is the main or sometimes the only K-idea, though this may be accidentally or deliberately reinforced by others. In spontaneous cases it may, in principle, be any idea common to both parties with which the transmitted idea X is in fact associated and which is in fact present to the recipient of the impression, etc.

initially present to A and associated in his mind with X was to be conceived as numerically identical with the K-idea presented to B (or some element or elements thereof), or only as closely similar. On account of the considerations adduced in this section I am inclined to the opinion that the first conception is the more reasonable of the two.

So far as I can see, the kind of view of mind in general and of individual minds which I have been trying to develop, provided we abstain from introducing artificial restrictions, etc., would at least permit the kind of interactions indicated above, and makes them seem quite natural. And the facts of telepathy are so difficult to explain at all convincingly in any other way as virtually to require them.

80. *Interaction of Minds, contd.* (3). But whether we rather laboriously start by thinking of individual minds as in principle isolated, and then invoke telepathy to show that they are not, or adopt the simpler course of saying that telepathy is a natural consequence of the fact that there was never any logical reason for postulating the isolation, we are, I think, driven to the same conclusion. That is to say, it is as gratuitous in theory and as inconsistent with observation to think of 'mind' as chopped up into a number of encapsulated units (individual minds) inherently incapable of interaction otherwise than through the physical mechanisms of speech, etc., as it is to think of the individual mind as a single indivisible entity. If there is one thing that modern psychology does make quite clear, it is that the 'individual mind' is not individual in the sense of being indivisible, but consists, according to the terminology adopted, of differing strata, sub-systems, groups of ideas, etc., often incompatible or even violently conflicting with each other. And if there is one thing which modern parapsychology[1] makes quite clear—in my judgement at least—it is that the 'individual mind' is not individual in the sense of being private, but is fundamentally linked with all other minds, with which it interacts in a kind of Federation of federations.

[1] Roughly definable for this purpose as the study of Telepathy, Clairvoyance, and more or less cognate phenomena.

I am, of course, speaking more of principle than of practice. In practice the mind of the normal person forms a reasonably coherent and fairly well-integrated system, and the influence of other systems (subject to certain speculative reservations to be made below) is usually negligibly small; but the principle involved, even though it may as a rule operate only feebly, seems to me to be as important to our conception of mind and mental phenomena as is the principle of universal gravitation to our conception of matter and material phenomena. The trajectory I describe as I fall over a cliff is but negligibly determined by the outer nebulae, the stars of our galaxy, the other planets of our system, or even by our nearest neighbour the moon; but if they were inherently incapable of exerting any influence at all—i.e. if gravitation were not omnipresent —I should not fall at all, and we could make no sense of the material universe.

Just as every particle of matter in the universe acts gravitationally on every other particle, though often in negligible degree, so, I maintain, every cognitum, cognitum group, or psychon acts on every other, so that any particular configuration of any system or sub-system is, in principle, the resultant of them all.[1]

Though not perhaps strictly on the main line of my discussion, it will be not without interest to explore briefly some of the more plausible implications suggested by the contention that the forces operative within the mind work also between minds and throughout the whole aggregate of cognita constituting mind in general.

In doing so I shall deal with various points which would usually be said to involve the conception of a 'common subconscious'; and I should like in passing to remove a minor misapprehension which seems to have arisen in connexion with my Association Theory of Telepathy. I have heard it suggested that, in order to explain telepathic phenomena, I am obliged to assume, or do in fact assume, the existence of a

[1] Cf. Hume on Association: 'Here is a kind of *attraction*, which in the mental world will be found to have as extraordinary effects as on the natural, and to show itself in as many and as various forms' (*Treatise*, Bk. I, Pt. I, Sect. 5; Everyman Edition, p. 21). (H. H. P.)

subconscious mental stratum or the like, common to all concerned. This is not correct, at least as I see it. The boot is rather on the other leg. The assuming is done by those (i.e. almost everybody) who take it for granted that so-called individual minds are necessarily or naturally insulated from each other. If one challenges this assumption, pointing out that there seems no *a priori* necessity for it, one will be told that, if it were not true, we should find other people's thoughts intruding into our minds, that this would result in endless confusion, and does not, in fact, occur. The answer, of course, is that it *does* occur, though not to the extent postulated by the objectors to make their case sound more convincing.[1] That is what we mean by saying that, after eliminating coincidence, etc., there are telepathic phenomena to be explained. But once the facts of telepathy are established, as they now are, the notion of a common subconscious follows from them; for this can only be defined, I think, by saying that it consists of such cognitum groups, etc., as are accessible to two or more minds, as ordinarily understood, to which it is common; or something very like this. And the essence of telepathy is that an idea known to be 'in' one mind appears in the field of consciousness of another, i.e. is accessible to both and 'common' to them. If at any moment it is not in the conscious field of either, then it is, by definition, in the subconscious, and constitutes, so far as it goes, a common subconscious. And if telepathy occurs on any considerable scale and between numerous minds, then the scope and extent of the common subconscious will be correspondingly enlarged. That is to say, the existence of a common subconscious is deducible from the facts of telepathy, on any reasonable definition of 'common subconscious', and would be so, I think, whatever explanation of telepathy were adopted. It is much more nearly correct to say that the common subconscious (more accurately *a* subconscious common to those minds concerned) follows from the facts of telepathy, than that it is necessary to

[1] Compare the case of a friend of mine (who really ought to have known better) seeking to discredit the possibility of Clairvoyance by derisively exclaiming, 'Clairvoyance—ridiculous! Telling me one can read all the books in the New York Library from Cambridge!'

assume it in order to explain the facts. But even this is misleading, and it is worth while examining the matter rather more closely.

81. *Conscious and Subconscious.* First, to what do we refer —what do we mean—when we use the word 'subconscious' either in an adjectival or substantival sense? Let us examine the facts. At any given moment there are certain cognitum groups, etc., which, being compresent with certain others (notably endosomatic sensa, 'vague mass of bodily feeling', etc.), constitute what we call my field of consciousness (and similarly, I am assuming, for anyone else). The field is not sharply defined, and is considerably fluctuating in content, as I have already explained. This, so far as I can see, is all we are entitled to call my 'conscious mind' at any moment. But there exists also a much larger aggregate of cognita which might form part of my field of consciousness at some other moment, but does not at the present one. For example, until a couple of seconds ago, the visual image of a certain house in Holland where I once lived did not, whereas it does now; it was readily accessible to my field of consciousness, as the facts show, and therefore indubitably a constituent of the psychon system forming what we call my mind, but not a constituent of my conscious mind. Speaking non-committally, we may say that it was part of my 'non-conscious' mind. There are innumerable other images, etc. (cognitum groups), which are equally accessible to my conscious mind (i.e. likely to come into it—I do not wish to suggest any active seeking), or substantially so; there are others which are much less so; and still others which are, as we say, 'repressed' and are exceedingly unlikely to appear at all in the field of consciousness considered,[1] or only under special circumstances such as psycho-analytic treatment. And the same applies, in varying degree and so forth, in the case of everyone else.

For certain purposes, notably psychiatrical, etc., it is convenient and profitable to dissect and analyse this mass of

[1] That is to say, my normal waking field as roughly defined above. They may appear, more or less openly in my dream fields, and may be permanently present in the fields of the complexes, etc., concerned, limited as these may be.

non-conscious material, to identify complexes and sentiments and study their interactions, or to distinguish various layers or strata. But it seems clear to me that the only natural and unequivocal distinction we legitimately make for such general discussion as the present is between those cognita which form part of So-and-so's momentary field of consciousness, and those which do not. The former constitute the 'conscious mind' at that moment, the latter the 'non-conscious'. Having made this point clear, I shall revert to the usual terminology and speak of 'the subconscious', or 'my subconscious', etc., as may be convenient, on the understanding that my use of the word does not involve any particular theories about the nature or organization, etc., of the cognitum groups referred to (unless expressly stated), other than that they are not constituents of the field of consciousness considered, i.e. the field of a normal waking person.

Now let us ask what meaning, if any, is to be attached to the statement that a given psychon or cognitum group said to be in my subconscious in the above sense (i.e. not in my momentary field of consciousness but quite likely to appear in it) is private to myself and is not, in the same sense, a constituent of your subconscious or that of anyone else. Up to a point, of course, the answer is easy enough. Memory images of events which have happened to me but not to you —including those of whatever elements of otherwise similar events were dissimilar in the various cases—obviously fall into the required category. My memories, for example, of landing a Shorthorn Maurice[1] in the sea near Littlehampton early in 1915, which are very reasonably vivid, are not normally accessible to any field of consciousness other than that I call my own, i.e. they are very unlikely to appear in any other.

But what are we to say about such experiences, or elements thereof, such as eating, which may fairly be said to be common to all mankind? No two instances of eating, of course, are precisely similar in all their details—I may eat peas with a

[1] Name of an almost prehistoric type of aeroplane. [The Maurice Farman 'Shorthorn' biplane of 1914–18. (H. H. P.)]

knife, and you with a spoon, and mine may be tough and yours tender—but unless they contain certain common elements they cannot properly be placed in the same class (we should not refer to them all as 'eating', though language is never a safe guide), and I think that everyone would agree that they can. And the same applies to innumerable other cases, such as the expansions of sentences referring to any public event, etc., of the material world—'The sun rose this morning', 'We live in houses', 'Fire burns', 'Dropped stones fall', etc. However much our individual experiences of these events may differ in detail, it will always, I think, be possible to analyse out *some* elements which are indistinguishably similar in all cases. That is to say, there are certain experiences, or elements of complex experiences which are usually and rightly said to be common to all mankind, or to nearly all, or to larger or smaller groups; but we would not usually say that the constitutive cognita are common to all members of the relevant group—we should ordinarily say that there are cognita (notably sensa) which, in each individual case, 'correspond to' or are 'caused by' the objects, events, etc.

But consider what I said in connexion with the gull-cry cognitum in an earlier passage. It is sheer gratuitous metaphysics, and quite illegitimate, to say that there are two (or more) different cognita unless they are distinguishable one from the other,[1] or are present simultaneously in the same field of consciousness; there is nothing whatever to prevent a single cognitum forming part of as many fields of consciousness as we like, any more than there is to prevent a man belonging to as many clubs as he can get himself elected to. To say that a cognitum must be two and not one, simply because it forms part of two fields, or two subconsciousnesses, is begging the question. Our cardinal principle is, roughly, that one may not make assertions ('These two entities are different') which it is inherently impossible to verify; and unless it can be shown, without question-begging, that two

[1] Perspectival differences, etc., must not of course be forgotten. When you and I see 'the same' view, there are inevitably perspectival differences between your cognitum and mine. Has the author made allowance for this in the next paragraph? (H. H. P.)

experiences which are obviously closely similar (e.g. two people looking at the moon from positions very close together) cannot contain indistinguishably similar elements, then we have no right to say that the relevant cognita are 'different'.

It seems clear to me, then, that the ordinary experiences of everyday life, if nothing else, must provide (or more properly consist of) a great mass of cognita which should logically be regarded as literally common to all experients; and in so far as these are not present in fields of consciousness, they will constitute, by definition, a common subconsciousness. Thus we would, in fact, be entitled to postulate a common subconscious in order to explain telepathy, if we wished; or we may equally well say that the facts of telepathy indicate the factual existence of it, as abstention from aprioristic restriction fully permits.

Whether we go in by the front door or the back we come to the same conclusion, namely, that Mind is fundamentally a unitary aggregate of interacting cognita, however close and dense and factually separated may be the condensations (individual minds) in it—much as the stellar universe is a unitary aggregate, though the condensations in it (galaxies, solar systems, etc.) may be close and dense and factually separated. There are no more intrinsically watertight compartments in the one than in the other.

I do not, however, like the practice of speaking of 'One Big Mind', or even 'Universal Consciousness', as adopted by some writers. Such phrases tend to suggest that the total aggregate is 'conscious' in the same sort of way as are the condensations we call individual minds, and equally 'conscious of', as we say, the kind of states we describe as 'feeling', 'willing', 'planning', etc.—but usually, by implication, more so. I see no justification for this, which I regard as exceedingly dangerous. I should be the last to attempt to deny, *a priori*, some kind or degree of consciousness to the aggregate as a whole; indeed, I think it a necessary consequence of the kind of view of consciousness I have tried to develop here. Personally, however, I should expect it to be vague, diffuse, and feeble, and not subject to any of these conditions; but this cannot be more than idle speculation till we have analysed more adequately the ways in which consciousness may vary and have correlated these with determinable factors.

82. *Group Minds.* Given that there is a mass of cognita to be regarded as a subconscious common to all incarnate men,[1] in so far as they partake of common experiences (and, indeed to other animate creatures also, so far as this is true of them) and certain other masses similarly common to particular groups, it is of interest to inquire what difference this will make to our conception of ourselves and the world. It is, of course, obvious that people similarly conditioned (i.e. subjected to substantially like experiences) will, on the whole tend to act and think more similarly than people dissimilarly conditioned; but I need hardly say that it is not this that I wish to discuss. It is a question of what *extra* effect, so to say, may be expected from the acceptance of this notion. And will such effects be discernible (if inherently not so, we shall be talking nonsense), and can they, in fact, be discerned?

The basic principle here is that any idea or psychon (cognitum group) common to two or more minds may act as a 'K' (see Section 79 above)—indeed, *must* act as a K, if my view be correct—for any other idea or ideas associated with it in any one of them. But this principle, if unqualified, soon leads to a kind of *reductio ad absurdum.* I eat my breakfast, and numerous thoughts (cognitum groups), notably those concerned with the work of the coming day, pass through my mind; you eat yours, and the same sort of thing occurs; millions of others eat theirs and think thoughts of their own. Now, in so far as there are indistinguishably similar cognita involved in these experiences of breakfast-eating, these are to be regarded as a single breakfast-eating set, so to say, common to all eaters of breakfast; and all these untold millions of assorted thoughts (cognitum groups) are associated more or less closely and more or less equally with this single set. They are, therefore, more or less equally accessible to the minds of

[1] I think we may omit consideration of discarnate 'spirits', human or otherwise, for the present purpose.

.ll the breakfast-eaters. The same applies to the thoughts vhich in each of us accompany and are, therefore, associated vith the performing of all more or less universal acts and the njoying of all more or less universal experiences. But the ffect of all this will be practically indistinguishable from nothing at all. To say that countless millions of assorted ideas are equally accessible to my field of consciousness, or equally iable to be called into it, so to say, by the breakfast-eating K's common to us all, is precisely equivalent to saying that the chance of any one of them appearing is vanishingly small.[1] My own 'thoughts' have an overwhelming priority, not because they have any magical quality of being 'mine', but because they are linked with that great number of other ideas (cognitum groups, psychons), images, etc., constituting the condensation known as my mind. Thus the psychical interactions, if I may use the phrase, due to the presence in the common subconscious of cognita constituting universal, or substantially universal, experiences, may be regarded as cancelling out and of virtually no effect. I shall, however, return to a different aspect of this question in Section 85 below.

This cancellation of effect would not, of course, apply in special circumstances where the supposed conditions were not fulfilled. If, for example, you and I happened, by some freak of chance, to be each presented with a python steak for breakfast—a far from universal experience, it may be presumed—and knew it to be such, not mistaking it for a bit of whale, then, if by virtue of peculiar experiences of your own, you happened to have formed a close association of pythons with the late Mr. William Wordsworth, the thought of that poet, I submit—or of Intimations of Immortality, or of the Lake District, etc.—would be more likely to come into my mind than they otherwise would be.

Or again and more importantly (since the above is no more than another example of how ordinary telepathy works), suppose that I were one of the few people having breakfast

[1] Unless, of course, the possible content of my field of consciousness were taken to be indefinitely large; whereas we know that, for one reason or another, it quite certainly is not, at least for all practicable purposes, even though its edges, so to say, may not be sharply defined.

at about the same time who had not just heard on the radio the news of some world-shaking event, such as the inadvertent destruction of Chicago by an atom bomb, or the unexpected demise of Dick Barton. In such a case I think I should still not expect any very specific single image or cognitum group to tend to appear in my field of consciousness. It is true that the breakfast cognita would be linked with innumerable cognitum groups constituting emotions of (we may suppose) rage, pain, and grief—or possibly, in some cases, of gratification or relief—but just in so far as these were indistinguishably similar they must be deemed a single set of cognita (as before) and would, therefore, be but one among many, and in so far as they differed the considerations just adduced would apply. And the same would be true of the various visual, auditory, etc., images constituting the ideas evoked by the news of the events in the minds of those hearing it. That is to say, if this reasoning be correct, I should be no more likely to have any kind of intimation that something unusual had occurred than I should if only very few people knew of it, despite the presence of the requisite K-ideas in the form of breakfast cognita.

This is of interest because it is a point that can, in principle, be tested by experiment—not quite in this form, of course, but by performing experiments in telepathy using varying numbers of agents[1] under otherwise similar conditions. Such experiments would be somewhat tricky to carry out, as there are a good many variables other than mere number to be eliminated, but that is irrelevant here. Very little work, so far as I know, has yet been done on this point, but one or two experimenters, notably Warcollier, have reported that a plurality of agents seems favourable to the effect; and this is, I think, what most people would expect on general grounds.

Now suppose that these indications were confirmed beyond reasonable doubt, as they at least well might be, what would then be our position? We could not jettison the principle of indistinguishability, for this is purely epistemological. I think

[1] i.e. persons acquainted with the 'idea' to be, colloquially speaking, 'transmitted'.

we should feel obliged to introduce the conception of a quantity akin to momentum, or inertia, and say that a cognitum which is a constituent of many minds has a greater inertia (and, therefore, in so far as the analogy is legitimate, a greater momentum and 'penetrative power') than one forming part of fewer minds, or something very like this. And if this be reasonable, then we might analogously expect to find that ideas which, by virtue of this property of quasi-inertia, came most easily into any given mind were also (I put it colloquially) hardest to get rid of. That is to say, an idea held by members of a community (sharing *ipso facto* a large number of K-ideas) might be much harder to displace from the mind of any one of them than could be accounted for by the physiological conditioning of the individual. And this is at any rate not discordant with observation. In ordinary parlance, people who are members of some political party, religious sect, or society of cranks do seem to hold their views with unnatural fanaticism and to be more impervious to reason than ordinary considerations would lead one to expect. But this is in the nature of a speculative digression within a digression itself somewhat speculative.

83. *Group Minds, contd.* But whatever be the value of the foregoing remarks, which are not intended to be more than suggestive, it is not this aspect of the subject that I wish mainly to consider here.

If we refrain from regarding individual minds as insulated from each other on *a priori* grounds (which, I insist, would be arbitrary and gratuitous), I think there is no doubt that they must be thought of as naturally tending to be linked in groups. We may slightly condense an obvious line of discussion by saying that ideas (cognitum groups) which are associated with the same idea are associated (more or less closely) with each other, i.e. will tend to accompany or closely follow each other in any field of consciousness more often than if they were not associated with that same idea. And this holds good, as all psychologists would agree, if for the words 'same idea' we substitute 'closely similar ideas', i.e. cognitum groups containing indistinguishable elements. Then assuming that the associative forces known to operate within minds operate also between them, it follows that if any plurality of minds, M_1, M_2, M_3, . . . etc., have any single idea, K, in common (or any set of similar ideas K_1, K_2, K_3, . . . etc.) then any other ideas, such as would ordinarily be said to be constituents of those minds, which are associated with this common idea K (or these K's), will be in some degree common to them all. That is to say, the K-associated ideas in any one of these minds will be more likely to come into the field of consciousness of any other of them than if there was not this K common to all. And note that any idea which has once done so, and has thereby become associatively linked with other constituents of the 'receiving' mind, so to term it, will itself become a K. Thus we should expect a kind of regenerative or snowball effect to occur, so that the communization of minds, so to call it, would take place more and more rapidly were it not for the action of dissociative forces.

I think it is worth while trying to get this clear by means of a simple example. Smith and Jones are, we will suppose, Sausage-worshippers, prostrating themselves daily before a string of these delicacies, frying them in blood on a golden pan and later consuming them in a ritual meal. In the course of these ceremonies they loudly proclaim (as would be but natural) the superiority of Sausage-worshippers over all other men, their divine right to impose their will on all consumers of such inferior substances as pies, pasties, or tripe, and their belief that the Salvation of Mankind depends on the universal dissemination of the Sausage Cult. Thereby their self-esteem is greatly flattered. Brown and Robinson, on the other hand, are mere sausage-eaters like ourselves, attaching to the comestibles nothing but the most ordinary dietetic and culinary significance. A well-fried sausage is to them a meal and nothing more. Both parties will, however, from time to time cognize certain groups of cognita, to wit, strings of sausages, containing many indistinguishable elements. These will act as K's, so that whenever Brown and Robinson encounter sausages they will tend to think not merely of frying-pans and gas-rings, and certain tastes and smells, but to some extent of blood and gold, and sausage-eaters' superiority and world domination. And the same process will take place, though in less degree, whenever they encounter blood or gold. And conversely, of course, whatever ideas *they* may happen to have associated with sausages, blood, gold, or superiority will tend to come into the minds of Smith and Jones whenever, for whatever reason, they happen to think of these things. But if the association of superiority, etc., with sausages, etc. is systematic and repetitive (as supposed) on the part of Smith and Jones, while the associations formed by Brown and Robinson are casual and haphazard, the former will gain the day and the cult of Sausagery will tend to spread to the minds of these latter innocents, subterraneously, as it were, through the common subconscious, without their at all realizing what is happening. This seems to me to be of great importance in connexion with the spread of religious and political creeds, often against (one would say) the natural proclivities of those affected by them. Substitute swastikas, Hitler-portraits, comic salutes, etc., or any other set of political or religious emblems, for the sausages and pans of the illustration, and mercy, tolerance, good faith, etc., or any other set of sentiments, for the feelings of superiority, etc., and it is easy to see how the process has worked, and how it might be made to work in a different way, as the churches perhaps have sometimes, if uncomprehendingly, been trying to do.

84. *Group Minds, contd.* There is one point I wish to stress in this connexion, which seems to me of considerable theoretical importance, namely, that the kind of effect just sketched will be the more important as what we would ordinarily call the content of the minds concerned is the smaller. In the extreme case, we may imagine two minds consisting of only two psychons each (which is absurd, but of no consequence here) of which one is common in the sense discussed above; that is to say, mind X consists of psychons A and K, and mind Y of psychons B and K (or K′ very similar to K). Then, if these be the only two minds in the universe, whenever K is present to X (A being necessarily present to

form a consciousness quorum, so to speak), then B must be present also, because there is no alternative; and conversely in the case of mind Y. So A, B, and K will always be compresent, and there will be only one mind, not two.

This is over-simplified and condensed to the verge of absurdity, but it should serve to indicate my point, namely, that the higher the proportion of K's common to two or more minds (as compared to the proportion of psychons not common to them) the greater will be the tendency for these other psychons themselves to become common and for the minds concerned to become unified into a single mind. This is relevant to the behaviour of crowds, where circumstances conspire to focus the attention of all members temporarily on a certain group of cognita (an orator, a prize-fight, etc.), i.e. to restrict the field of consciousness of each to a particular aggregate of cognitum groups, all acting as K's.

But it is perhaps especially of interest in connexion with animals and their instincts.[1] The private and individual experiences of an animal of any species are evidently very limited compared with those of any human being, partly because of their restricted range of effective travel,[2] more to the lack of hands, but above all to the absence of speech and writing, whereby we enormously extend our range of experience to include all manner of experience by proxy. Thus the psychon systems of animals in general, and of members of any given species in particular, will contain a far higher proportion of K's than those of even primitive man, let alone the sophisticated and highly individualized products of to-day; and this will be the more true, broadly speaking, the lower in the evolutionary scale we go. Thus, it seems to me reasonable to think of any given species, cats, say, as possessing a collective or group mind of relatively large proportions, consisting

[1] Cf. the author's *Telepathy*, pp. 159–60.

[2] Some migrants, of course, travel great distances, but this does not of itself add much to the variety of their experience, which is a matter of difference of environment rather than of geographical distance. [One would suppose that the summer and winter environments of e.g. a swallow were very different indeed. But perhaps the bird does not notice the differences as a human traveller would? (H. H. P.)]

of the cognitum groups common to all cats, and exhibiting relatively small differences between the individual 'excrescences' (sub-condensations). For dogs, which, though I am a cat-lover myself, I must admit to be definitely more intelligent, the individual differences will be greater; while, near the other end of the scale, earwigs will have virtually nothing but an Earwig mind, with next to no individuality at all. And this corresponds exactly with what I might term the distribution of instinct, all the way from amoeba up to man. At the lowest levels behaviour—if one can fairly call it so—is purely instinctive and merges indeed into mere physico-chemical reaction; but the higher up the scale one goes, the more it is modified by individual peculiarities, till in sophisticated adult man it operates directly only in respect of the most fundamental activities of self-preservation and reproduction (and not always there, as records of heroism and asceticism show), though still, of course, importantly influencing both thought and conduct.

Now, if we omit to make the customary assumption, for which there is no *a priori* justification whatever, that the existence of cognitum groups and sequences organized in one fashion (minds) is dependent on their relation to cognitum groups organized in another fashion (material objects, bodies, brains), we reach this rather interesting conclusion. There is no *a priori* reason why the psychon system I have suggested as constituting the collective mind of a given species should not comprise all the experiences (cognitum sequences) of all members of that species, past as well as present. Thus, any member of the species has, so to say, access to and can draw upon the accumulated experience of the whole species throughout its history. Thus, on this view, when the garden spider spins her web, the tailor bird stitches leaves together to make her nest, or the beaver builds his lodges and his dams, they are not acting out of private intelligence, and it is not necessary to suppose (which, frankly, I have never found credible) that all these intricate and beautifully co-ordinated movements involved in these activities are determined by a correspondingly intricate pattern of nerve-cells, synapses, etc.

in the brain, inherited by virtue of the chemical constitution of a molecule in a gene. The creatures are guided by the accumulated memories of the species, much as I am guided by my accumulated memories of parcel-tying whenever I tie up a fresh parcel. This may sound a trifle fantastic to those unaccustomed to thinking in these terms, but to me at least it seems a good deal less so than the only alternative so far presented to us.

85. *Mind as a Whole. The Group Mind of Humanity.* Much of the foregoing may well have appeared to the reader to be somewhat irrelevant to the main line of our inquiry; and so perhaps it is, but I have been trying, by presenting snapshots taken from different angles, to invest with some degree of solidity and coherence the flat statement that Mind in general and so-called individual minds in particular are neither more nor less than assemblages of cognita organized in patterns other than those characteristic of material objects. I have tried to show that, if we take this view and refuse to complicate it with arbitrary apriorisms, we may reach certain provisional conclusions which appear to be well in accordance with observable facts. In particular, I have stressed the importance of not arbitrarily supposing that there is one set of 'laws' which holds good within minds, and another set which holds between minds; but of adopting, rather, the much more natural hypothesis that there is just one set of laws which holds throughout the whole mass of mentally organized cognita—as is indeed no more than tautological.

I shall now try, as best I may, to pull the whole picture together, and for the sake of simplicity I shall confine myself to the minds or Mind of humanity, leaving the animals, etc., to take care of themselves, though I do not believe that there is any more hard-and-fast line to be drawn between them and ourselves than anywhere else.

If we work out as best we can in the light of our present knowledge the kind of consequences likely to follow from the interplay of associative and dissociative forces, I think the most important conclusion we reach is that there is no essential difference of kind, but only of magnitude and degree

—and, one may put it, of 'happenstance'—between the mind of the individual and the mind of humanity as a whole. Individual minds, it is true, are to a great extent insulated from each other in practice; but so are the complexes, sentiments, and multiple personalities which are to be found, invariably, frequently, or occasionally as the case may be within what we regard as a single mind. The differences of separation and fusion are differences in degree, not of principle. That is to say, the difference between my mind and my fellow man's is the same *kind* of difference, and arises in the same kind of way, as the differences between two members of a pair or set of multiple personalities; but whether we are dealing with different people, in the ordinary sense, or with different personalities in a pathological case, or with different moods, etc., in a normal person, there is always what may fairly be called a common subconscious if only we go deep enough to find it. To use a simile[1] which is valid up to a point, we may think of psychon-systems, sub-systems, sub-sub-systems, etc., i.e. of mind in general, as a congeries of aggregates of assemblages of archipelagos. The individual islands will be cognitum groups (psychons), a group of such islands will be a complex, sentiment, or the like, an assemblage of such groups will be a so-called individual mind, and aggregate of such assemblages will be, perhaps, the collective mind of a nation or other society, and so on. The bits of land actually projecting above the water will be conscious fields (roughly—the analogy becomes rather weak here). The common subconscious will be the masses of rock, etc., below the surface, and the same type of mechanical forces will operate therein as on the islands themselves. Assuming the islands to be the peaks of submarine mountain ranges, and not the tops of perpendicular pillars, the islands of any small group will, in general, be connected at a lower depth than those of different groups, and the groups of an assemblage at a lower depth than those of different assemblages, and so on.

But note importantly here that there may be any amount of criss-cross ridges connecting group with group, and

[1] [First suggested, I think, by F. W. H. Myers (H. H. P.).]

assemblage with assemblage, etc., so that the islands might be found to form very differently connected patterns if we were gradually to drain the water away. Look at the contour lines of any hilly district on the map, and note what very different arrangements would be formed by flooding the country up to the 100-foot, 200-foot, 300-foot, etc., levels.

Except for the facts that the illustration is both static and spatially dimensional, it serves the purpose very fairly well.

Take the matter now from another point of view. All men, we may suppose, and presumably nearly all animals, enjoy experiences (cognitum groups) of breathing, eating, drinking, sensations of cold or warmth, reproductive impulses, etc., etc. These, one might say—with whatever others are similarly universal—form *the* common subconscious. Nearly all men also experience sensations of light and shade, seeing the sun, moon, and stars, the properties of certain material objects and substances, etc.; and these will be part of the common subconscious of most human beings. Such virtually universal experiences will serve as K's; but, just in so far as they are virtually universal, and therefore indiscriminately linked with an innumerable multiplicity of more or less 'private' and individual experiences, their effects will cancel out in the manner already indicated.[1]

But above this level, or thereabouts, processes of interaction and group-formation will begin to set in. Ordinary conditioning apart, any plurality of persons having a number of cognitum groups in common, but peculiar to that plurality, will tend to be more 'like minded' and to form a group-mind more readily than those who have not. Thus, the inhabitants of any particular type of country (flat, mountainous, snowy, arid, fertile, etc.), or speaking the same language, or following the same religion, will *ipso facto* form a group-mind—up to a point—in virtue of their common (but peculiar) cognitum groups acting as K's. And the same will be true, of course, of all who follow a given profession, or are born in a given social class, etc., etc.

These groupings, however, will be interwoven in an

[1] Cf. Sect. 82, p. 215, above.

exceedingly complex way. All German-speaking, French-speaking, Italian-speaking, and Romansh-speaking Swiss, for example, will tend to form groups on account of the K's provided by the languages, but to be federated in a larger group by those representing the political history and natural features of their country. But all French-speakers will tend to federate together, and so will all doctors or lawyers, all Protestants, red-headed men, vegetarians, believers in astrology, diabetics, nonagenarians—and so on almost indefinitely.

I do not wish to suggest that the effects of these interrelations and cross-groupings will necessarily be other than very slight, or (perhaps) in practice negligible. In the first place, even in the kind of case imagined, of two Protestant, nonagenarian, diabetic, French-speaking, etc., Swiss doctors, the number of K's involved may be small compared with the total mass of differentiating experiences. In the second, and more importantly, the possession of numerous K's alone is not sufficient to ensure or greatly facilitate the sharing (or 'transference', as we loosely say) of other ideas between the two or more minds concerned. Nothing will happen unless the idea in question is associated with the K's in one mind or the other; and if it is only associated with, say, one of them, the K's with which it is not associated might as well not be there.[1]

But I do contend that, unless we understand the kind of way in which minds are built up and interrelated—and it seems to me that the *kind* of view I have put forward ('errors and omissions excepted') is the only rational one—then we have very little chance of working out a sound psycho-sociological science, such as we so desperately need at the present time.

[1] This again is a point open to experimental test, at least I think so. I I am right, then increasing the number of K's with which an idea to be 'telepathed' is directly and specifically associated by the conditions of the experiment will be more likely to promote good results than merely picking as apents and subjects people whom we think likely to be 'like-minded' on general grounds, e.g. homozygous ('identical') twins, as has been suggested. But, of course, the experiences common to these might act indirectly to an extent at present unknown.

It may not, on the face of it, make very much difference whether we suppose the celestial bodies to move as they do because they are pushed around by intrinsically autonomous angels, who co-ordinate their movements by flag-wagging across the firmament, or whether we adopt the Newton-Einstein gravitational scheme; but there can be little doubt as to which is the sounder foundation for astronomical theory. It may be more trouble to learn and apply the methods of the tensor calculus than to argue about what will happen from the assumed psychology of angels, but we are very much more likely to obtain reliable results.

Similarly, it is less trouble and probably more fun to talk about Egos or Souls or other metaphysical unobservables, and make unverifiable deductions from their alleged properties, than to tackle the formidable task of devising and applying a logico-mathematical technique for dealing with the groupings and interactions of cognitum systems, but again there can be no doubt as to which is the firmer foundation to build on. I doubt whether we yet know even how to set about the task, though a first-class mathematical physicist would probably not take long to set us on the right track. But however long it may take, the work has got to be done, if humanity is ever to develop some measure of sanity (roughly, correspondence between thought and fact) in its social relationships.

86. *Concluding Remarks.* Perhaps the most important thing to be said about this long, and I fear somewhat wearisome, chapter is that, strictly speaking, most of it need not have been written at all. There are only two or three points which are really vital to the main line of my discussion; all the rest is in the nature of illustration, emphasis, and tentative working-out—with little of which do I feel any great satisfaction. The points which do seem to me to be important are these:

First: as always, our guiding principle must be that it is futile and worse to make assertions the truth of which cannot possibly be tested. If we do, then we might just as well make the opposite or contradictory assertion and shall only have wasted our time.

Second: this implies that any assertion we make or hypothesis we advance must be about cognitum groups and sequences, or reducible to statements about these; for, if it be not, then it is necessarily concerned with inherently unobservable and unplaceable pseudo-entities, and must be rejected as meaningless under the principle just mentioned. We must, therefore, refrain from mentioning Egos, Selves, Spirits, Souls, Wills, and all like metaphysicalities—except, of course, in so far as (understanding what we are doing) we may find it convenient to use the *words* as shorthand symbols, capable of expansion, for certain types of cognitum group, sequence, system, etc.

Third: for precisely the same reasons, we must abstain from making *a priori* assumptions about the supposedly inherent and necessary limitations, restrictions, isolation, etc., of minds or cognitum groups, etc., generally. In particular we have no *a priori* right to declare that the 'forces' which we observe to operate within minds do not operate between them.[1] Whatever limitations, etc., there may be, or whether forces do or do not operate between minds, are matters to be settled by factual observation, not by *a priori* assumption, however tacit.

Fourth: in studying the individual mind we do, in fact, observe certain phenomena of association and dissociation, which we may conveniently speak of as due to corresponding 'forces'. In accordance with the foregoing principles, therefore, we may—and I think must—suppose that these forces operate between minds just as they do within minds—*en gros* as they do *en détail*, so to say. In this we at least follow the example of Newton (though, of course, eminence of precedent is not of itself a guarantee of correctness), whose supreme intellectual achievement was, I suppose, that of realizing that the movements of celestial bodies could be ascribed to the operation of the same all-pervading force as the fall of the famous apple.

[1] But note what I said about 'forces' above. [Sect. 67, p. 165.] Perhaps this could be more accurately expressed by speaking of 'occurrences which we find it convenient to describe by saying that they are due to the operation of forces', etc.

Once we have got this far we have done all that is fundamentally necessary for our basic understanding of Mind. All the rest is a matter of working out detail and filling up outlines in the light of such knowledge as we have of how these forces do in fact work (corresponding to observations on how planets do in fact move and bodies fall, etc.). And for the purposes of my main argument it does not in the least matter whether such tentative attempts as I have myself made have been well or ill conducted. Wherever I have gone wrong, as probably all too often, and whether as a matter of what I have taken as observed fact or in reasoning therefrom, someone else will in due course get it right. But the sooner the better.

I do not propose, therefore, to summarize this chapter in greater detail here, but will pass on to the question of the actual relation and possibilities of interaction of Matter and Mind, which is, in a sense, the core of this book.

VI

MIND AND MATTER

What is mind?—no matter
What is matter?—never mind.

Popular saying

87. *General.* There seems no reason why this chapter should not be mercifully brief. The basic answer to the question of how Mind and Matter are related, and, in principle therefore, of how they may interact, was given at the end of Chapter IV.

Words such as 'Matter', 'material object', or 'material event' refer to cognitum groups and sequences which follow each other in a certain type of pattern, or with a certain kind of regularity, namely, the type codified as what we call the laws of (material) physics. Words such as 'Mind', 'mental object', or 'mental event' are cognitum sequences which follow a different sort of pattern, to wit those codified in the laws of psychology, so far as these are at present understood. There is no intrinsic difference between the cognita constitutive of the one type of sequential pattern and those constitutive of the other. The differences between the referents of the two sets of terms are solely differences of organization, not of the nature of the constituents.

There is, therefore, no difficulty of *principle* (however much there may be in working out detail) about understanding how mental and material cognitum systems may interact and influence one another, just as there is no difficulty in understanding how Greeks organized in line of phalanxes can interact with Romans organized in echelon of cohorts, however difficult it might be to forecast the precise course and outcome of the conflict.

The whole trouble has arisen from the uncritical use of such terms as 'Substances' or Things-in-themselves *having* attributes, or Minds-themselves, Egos, Souls, etc., which *have* properties, faculties, and so forth. This has created a wholly unnecessary set of difficulties, which have puzzled philosophers from time immemorial, and need never have arisen if only they had insisted on a ruthless positivism and been content to examine what is actually going on.

88. *Classical views.* It would obviously be waste of time to make any sort of detailed study of views which I hold to be almost without exception meaningless and nonsensical, but there are a few points of interest worth noting.

All such theories fall of necessity into one or other of two main groups, Dualistic or Monistic.[1] The Dualists take it for granted that there are two radically different types of existent or substance, Matter and Mind, and their problem is, of course, how Substances which are radically different by definition, none the less contrive to interact, as they manifestly do—in the sense that there is an undeniable correlation between what are commonly called physical events, such as being hit in the eye and what are commonly called mental events, such as feeling angry.

The Monists, on the other hand, who contend there is only one Substance, divide as to whether Everything is Matter, or Everything is Mind—though one would think that if Everything is one substance, to wit X, it does not make much difference what pet name we give to X; whether we call it Matter or whether we call it Mind, it will be none the less or the more 'the' Substance that is Everything. At rock bottom I cannot see any way of subjecting the two alternative contentions to empirical test, and I think the distinction may accordingly be dismissed as meaningless.

Suppose you start out with a perfectly open mind, examine all the relevant phenomena and come to the conclusion that All is Matter; and that I, supposedly of equal intelligence, examine the same phenomena with equal fairness, and conclude that All is Mind. We have arrived, from the same data, at diametrically opposite results (or are they?), except that

[1] For a clear and readable account of this topic see Dr. Gardner Murphy's article (*Bull. Amer.* S.P.R.). [*Journal* of the American Society for Psychical Research, October 1945? (H. H. P.)]

we are agreed that 'All is Something'. How are we to decide which of us is right, or whether there is anything but a name ('Mind'-'Matter') between us at all? 'All is X' seems so very equivalent to 'All is X″' unless we are clear to start with as to how X and X′ differ; but this we cannot be, as I have been at pains to show, since neither of the alternatives ('Matter', 'Mind') can have any meaning *per se* at all.

It would hardly appear, therefore, the taste for miraculous intervention being by now thoroughly outmoded, as if any of these types of theory were likely to prove a sound foundation for constructive thought.[1]

89. *Epiphenomenalism: Neutral Monism.* I may be ignorant, but I suspect that the difficulties are (supposedly) dealt with in the minds of those concerned mainly by a lavish use of the 'blessed word' *Epiphenomenalism* (I do not know to whom this is due[2]). This seems to me to mean no more but to play no less a part than the Divine Will found so valuable by Descartes and his followers. Nearly all orthodox scientists of to-day are, I think, devout materialists (Monists) at heart; but when confronted with specific phenomena of consciousness occurring in themselves, they take refuge in saying that although such phenomena are no doubt 'real' enough in their way, they are 'merely epiphenomenal' to those of physics, and imagine that they are thereby answering the question or at least are saying something of relevance. I may be very stupid, but so far I have never been able to ascertain what, if anything, a scientist means when he uses this term. Nor, I suspect, has he.

The doctrine of Neutral Monism, which, as readers will have noticed, comes so very close to my own in so many respects, should properly receive more careful consideration than I can give it here; and I should be only too pleased to accept it in its entirety if only I could persuade myself that it is substantially identical with mine or that it could enable us to understand an equally wide range of problems.

Mr. Bertrand Russell's *Analysis of Mind*, referred to at the

[1] Cf. Ch. IV, Sect. 56, p. 140 (H. H. P.)
[2] To T. H. Huxley, I believe. Cf. above, p. 9. (H. H. P.)

end of Chapter IV,[1] comes very near indeed to saying what the basic relation between mind and matter really is, so far as it is possible to provide an exact answer to so vague a question; but it does not seem to me that the distinguished author has fully rid his mind of the supposition that the words 'Matter' (generically) and 'Mind' (generically) must refer to existents of some sort in the universe and that there is a kind of matter-mind or mind-matter substance behind both types of objects, etc., *out of* which both are 'formed' and the nature of which could conceivably be the subject of further inquiry. But if so, it is only a Metaphysical Unobservable; and as such it is equally anathema whether one arrives at it at the end of a discussion, or postulates it as a starting-point.

[1] p. 140, above.

APPENDICES

I

DON'T SHOOT THE PHILOSOPHERS—YET[1]

It would be a mistake to suppose that, because philosophers talk nonsense, their influence is of negligible importance in world affairs, and may safely be ignored; that, when they do harm, as sometimes happens, they do it on purpose; or that, because their remarks are for the most part unintelligible it is impossible to deal with them.

But first: Do philosophers talk nonsense? Not necessarily more than other men, and probably less, if by 'philosophers', you mean only 'lovers of wisdom', 'men of inquiring mind', or the like. But if you use the term in the technical sense of those whose main activities are the study of pure logic and metaphysics, the answer is an unqualified 'Yes'. Pure logic is all right, because it does not claim to be more than a tool (like mathematics, which is strictly a branch of it); and it has, of course, been of immense service. But metaphysics, the characteristic philosophical activity, is another story, and it is gradually becoming recognized that all metaphysics not only is nonsense, but inevitably so, and cannot possibly have anything to say about the actual world.

The metaphysician is one who attempts to draw conclusions about matters of fact by arguing from certain assumptions, premises, or 'axioms' which he claims to be 'self-evidently' true (or true *a priori*, as the phrase is). Note that this enables him to 'prove' any conclusion he wishes, if only he works backwards with sufficient cunning to find a set of premises from which the desired conclusion will follow, and which he can persuade you to accept. And, of course, *if* his premises are true, and his logic correct (as it usually is), his conclusions must be true also. Unfortunately, however, no premise can *self-evidently* be true unless it is what is called 'tautological',

[1] Cf. Ch. II, 'The failure of metaphysics', above. (H. H. P.)

i.e. is concerned only with the use of words (or other symbols); in this case its truth may be assured by definition, but it cannot have any bearing on matters of fact. If, on the other hand, it is concerned with facts, then conclusions about other matters of fact may be deduced from it; but in this case it will itself require verification. If a man proceeds on these lines from observable fact, by deduction, to other observable facts, then he is not a metaphysician but an empirical scientist. The, metaphysician can only preserve a domain of his own by making statements which sound as if they were about matters of fact (e.g. 'the Absolute has no Qualities') but are inherently unverifiable, and impossible, therefore, to regard as either true or false, i.e. are nonsense. This is the first philosophic fallacy.

The second is the naïve belief that, because there is a word in the language, there *must* be a 'something' in the actual world to correspond with it. This is simply untrue. Putting it rather briefly, to use such words as, for example, Reality, Essence, Substance, Continuant, Absolute, is either cheating or else implies that there are existing entities in the universe to which these words refer; but to verify such implied statements we must deduce the consequences that would follow from the existence or the non-existence of the entity in question, and make observations of some kind to ascertain whether these consequences occur or not; but this procedure the metaphysician cannot even prescribe, so that is is impossible to tell whether these alleged entities exist, and the words are accordingly meaningless.

The third fallacy is like unto the second, and consists in supposing that grammatical form necessarily corresponds to the structure of the universe, which can therefore be deduced from it. For example, noting that the statement 'Jones has a gold watch' is identical in grammatical form with 'A lemon has an acid taste', the philosopher usually infers that there must be a 'somewhat' which stands in the same relation to the acid taste as Jones does to the gold watch. This again is nonsense, for this alleged 'somewhat' (thing-in-itself) cannot conceivably be observed or its existence verified; for by no process whatever of studying a lemon can one possibly observe anything other than more properties, qualities, etc., such as are commonly said to be 'of' the lemon. The error is of purely linguistic origin, but it has led to an infinity of trouble, as have many others like it.

Such, in briefest outline, is the case for contending that all philosophers, when speaking as metaphysicians, invariably talk meaningless non-sense.

Now for their influence. To take two outstanding examples: the two most important factors in world affairs in the last fifty years have been, I suppose, German Nazism and Russian Communism. For some reason which I do not profess fully to understand, the sponsors of Communism (which is essentially a theory about the best way of producing and distributing goods, to be tested by finding out how well it works in practice) have seen fit to turn it into a religion, and to support it by an elaborate philosophic system known as Dialectical Materialism. This system derives from Hegel, through Feuerbach, Marx, Engels, and Lenin. But unfortunately Hegel probably talked more, and more childish nonsense than any philosopher who ever lived,[1] so that it is not surprising that we find Communist doctrine crammed with ridiculous phrases like 'identity in difference', 'individual universal', 'interpenetration of opposites', etc. I do not say that there would have been no revolution in Russia without this system—there obviously would—or that Communism would not have been tried (the basic idea of common ownership is no new one); but I doubt whether the attitude of irrational fanaticism, intransigent cocksureness, and unassuageable suspicion, which makes the Russians so difficult to deal with, could have been created without this slogan-yielding creed.

The doctrines of German Nazism may fairly be said to have started with Fichte's *Addresses to the German People* of 1807, in which their superiority to all others was affirmed (on the ground, at this stage, of 'purity of language'), and the

[1] Cf. Russell, *Our Knowledge of the External World*, first edition, pp. 38–9.

necessity for moulding them into a 'corporate body', with universal military service, no individual will, etc., etc., was proclaimed.[1] The work was carried on by Nietzsche[2] (with all the rubbish about 'Superman', 'the 'big blond beast', etc., and all the emphasis on Will—whatever that may mean, which is not clear) and enthusiastically supported from various quarters—not only German—till it culminated in the ravings of Hitler, Rosenberg, and Streicher.

In both cases, of course, the philosophers could not have produced the effect they did if there had not been predisposing tendencies towards the acceptance of their views in the peoples they addressed. This is always the case, for good as well as evil; for men do not, in general, accept views or adopt beliefs because they have previously been shown to be true; they adopt the beliefs first because they want to believe them (e.g. that they are members of an innately superior race, etc.) and then cast around for arguments with which to support them. And if a philosopher provides an apparently impeccable system ready made to measure, it is naturally seized on with avidity.

But perhaps the most influential—indeed, most pernicious—metaphysicians of the lot have been the physical scientists, who do not primarily profess to be philosophers at all, but whose pronouncements about the nature of the universe are backed by all the enormous prestige of their material achievements. So long as they confine themselves to co-ordinating their observations, and expressing the observed regularities as 'Laws of Nature', we have nothing but praise for their prowess. But when they step out of their province and declare that 'Matter is the only Reality', or 'It is impossible for Mind to exist apart from Body', or the like, then they fall instantly into the same traps as the philosophers, and equally talk non-sense. Their trouble is that they have been brought up in the tradition of the classical metaphysic, with these mythological 'things' somewhere behind, or within,

[1] See Russell's article, 'The Ancestry of Fascism' (1935), reprinted in *Let the People Think* (Watts, London, 1941).

[2] Nietzsche, I think, whatever his faults may have been, was neither a Totalitarian nor a Pan-German. (H. H. P.)

or beyond, or above the 'appearances' observed. Consequently they have been unable fully to realize that they are never observing 'matter' (except in the sense that this is a convenient shorthand term for what they actually study), but only the coloured and shaded patches, feelings of touch, warmth, pressure, taste, smell, etc., commonly said to be appearances 'of' it; or that the difference between 'Mind' and 'Matter' is not one of fundamental constitution but of the patterns, so to say, in which these directly observable entities are arranged. Following one pattern (laws of physics) these observables *are* what we call 'matter'; following another (laws of psychology) they *are* what we call 'mind'. But we know little about psychological laws and much about physical, so it is easy for the physicist to conclude that what he understands and can cope with is alone 'real'. Hence, with the aid of but little laziness and wish-thinking, arises crude Materialism with its concentration on material ends and immediate profit, to the exclusion of all other considerations. Which is the bane of the world.

But the philosophers are not to be blamed for all this muddle; they mean no harm. It is not a matter of malice aforethought, or of congenital stupidity, for they are, in the main, the most amiable and ablest of men. It is due to two main causes: first, to uncritically accepting the naïve idea of the omnipotence of reasoning from *axioms*, which is an inheritance from the ancient Greeks, who applied the method with triumphant success to geometry (because their axioms happened to be very nearly true of the space in which we actually live); second, to the equally uncritical acceptance of the superficial implications of linguistic forms and usages—that is to say, to not first studying the tools of their trade, and, in particular, to not working out a satisfactory theory of *Meaning*. But it would have required almost praeterhuman foresight to do so.

Now, the way to counteract views which seem pernicious (and incidentally to clarify one's own) is not to meet the philosophers on their own ground and argue against them—they are far better at that than you are—but, remembering what I have just said, to take either or both of two lines. First, insist on being told how they propose to verify their assumptions: if they reply that they need no verification, because they are 'self-evident', retort with a polite sneer, 'Oh. Mere tautologies. You can't get anything about the actual world out of *them*.' Second, insist on being told what they *mean* by the words they use, and do not be content with a mere dictionary answer—a mere substitution, that is to say, of one set of words for another. Philosophers must always be chivvied from between the covers of the dictionary, out of the world of words and into the world of observable fact; if they cannot, sooner or later, define their words in terms of possible observations, those words are meaningless, and their discussion just so much waste of breath. And if you want real fun, try the effect of asking a philosopher what he means by 'meaning'. If he replies otherwise than to the effect that the meaning of a word is, directly and in the first instance, the psychological state which causes its utterance, or which its utterance causes, in speaker or hearer respectively, and thence, indirectly, the entity to which it would ordinarily be said to refer—then he does not know his job, or what happens when he talks.

Only when thinkers generally, and philosophers in particular, have freed themselves from these errors, and have brought their untrammelled minds to bear on the age-old problems of human life and destiny, will there be a chance of working out that rational philosophy, or religion, or theory of living, which the world so desperately lacks at the present time.

Meanwhile, *please* don't shoot the philosophers yet; they are doing their best, and they have such pretty ways.

II

LIFE AFTER DEATH

THE NEED FOR AN INVERSION OF THOUGHT

The late F. W. H. Myers, the great pioneer of scientific work on this subject, used to tell an agreeable story about how he once asked a man, at a dinner table, what he thought would happen to him when he died. At first the man tried to evade the question, but when Myers pressed him he replied, 'Well, I suppose I shall inherit Eternal Bliss—but I do wish you wouldn't talk about such unpleasant subjects!'

I suppose that, apart from a few specialists like myself, the more fervent spiritualists, and a comparatively small number of people who take the religious pronouncements of different eras quite literally, this is very much the attitude of most of us to-day. Intellectually, we all realize that we shall not live, physically speaking, for ever; but, on the whole, we prefer to think as little as possible about what will happen next. And this is only natural, because our basic biological instincts vehemently impel us to seek life and evade death at almost any cost, so that there is bound to be a tremendous emotional resistance to even the thought of dying. None the less, the matter is of some importance, though not one with which it would be wise to preoccupy ourselves; and it is not a bad thing to take stock occasionally of what positive progress, if any, is being made.

Unfortunately, the mere fact that the subject is of such intense emotional significance—if not on our own account, as we may stoically flatter ourselves, then on account of others in cases of bereavement—imposes the heaviest of handicaps on clear thinking; and we are far too apt to accept uncritically any argument or alleged evidence which either nourishes our secret hopes, or, at worst, enables us to shelve the whole problem as insoluble and one about which we cannot reasonably be expected to do anything.

The consequent confusion of thought has to be studied to be believed, and ranges from the philosophers who imagine that they can prove 'The Immortality of the Soul' by deductive argument from 'first principles' or 'axioms', of which the truth is guaranteed by 'intuitive certainty', and without any reference to observable fact—which is flatly impossible—to enthusiasts who take every word at a spiritualist séance at its face value—which is merely silly.

The first and perhaps most serious source of this confusion lies in the uncritical but almost universal assumption that we know to start with what such words as 'soul' and 'survival' *mean*, i.e. that they refer respectively to an entity or a state of known character, previously given in experience, so to say, so that there is nothing ambiguous about the question whether a man survives death or not.

This is simply untrue. We do *not* know what is meant by the statement 'Smith has survived death' in the same sense that we know what is meant by 'Smith has survived shipwreck'; and until we clear our minds on this all-important point it is no use wrangling about whether survival is a fact or not.

Consider: there is no doubt what we mean by 'Smith's body', for we can give precise instructions for defining and identifying it by observations and measurements. If these have been made during Smith's lifetime, and a body is found after the shipwreck, there is no difficulty of principle in deciding whether it is Smith's, or in saying whether it is alive or dead, according to how it responds to certain tests. But how are we to define or identify 'Smith's soul' or decide whether *it* has survived? It is no use saying that it is 'that part of Smith which survives death', or 'is capable of existing apart from the body', or that it is 'an immaterial essence', or the like; for to do so would be merely to beg the whole question. And if we leave it undefined, then we quite literally do not know what we are talking about, and our discussion is likely to be futile.

The only thing we can do, and remain rational, is to adopt the same kind of process as for Smith's body. We make certain observations on Smith's physical characteristics, and, for purposes of identification, these *are* Smith's body; similarly, we must make observations, or have data available, on Smith's 'mental' characteristics—notably memories, but also perhaps intellectual abilities, etc.—and, for purposes of identification, these *are* Smith's 'mind' (or 'soul' or 'spirit' or what you will). For the moment we try to go beyond observables, we are instantly sunk in the morasses of the meaningless.

It is true that, in practice, these 'mental' data consist of reactions (words, etc.) given by Smith's body—and we usually take it for granted that, in so far as they are characteristic and peculiar, they cannot be obtained from any body other than Smith's. But this is a matter of habit and inference, not of logical necessity. What is logically inescapable is that, unless we define Smith's mind in terms of actual observations such as these, we cannot talk sense at all about whether it has survived or not. And if, after the death of Smith's body, we do in fact find the same (or significantly similar) reactions to our tests (e.g. in reply to questions) in connexion with another body[1]—or even, theoretically, a machine—and cannot account for this in any other way, then we not only may but must conclude that Smith's mind has survived; for there is no other rational meaning to be attached to the statement. Smith's mind can only be defined in terms of a group of observed processes, etc.; and if these survive, Smith's mind survives. That is to say, the whole question is not one for metaphysical speculations, all of which are meaningless from the start, but for the most objective methods of physical science.

The second great error is to suppose that reactions of this kind (using the term in the widest sense) are not in fact obtained, i.e. that there is no worthwhile evidence for survival. This again is simply not true, for there is abundant evidence of very high quality. We must, to be sure, set aside most of the material relied upon by the more clamorous spiritualists, because these amiable but uncritical people have extensively

[1] e.g. the vocal organs of an entranced medium. (H. H. P.)

discredited their own case by their omnivorous acceptance of every alleged marvel proffered by charlatans, and by their stubborn defending of the indefensible. But anyone who takes the trouble to study the material collected by the leading Societies for Psychical Research, through mediums of the calibre of Mrs. Piper, Mrs. Leonard, and Mrs. Garrett, and through private automatists of the highest integrity, must conclude that the case for at least some sort of survival (admittedly perhaps only partial) is much stronger than any court of law would demand in a suit by a missing heir to a fortune.

To give examples is, perhaps, risky; it is rather too like saying, 'There is plenty of money in England, here is sixpence to prove it'; but the following three cases, taken almost literally at random, will serve to illustrate.

(*a*) My cousin, D. G. H., to whom I was much attached, was killed in April 1915 in the First World War. In December 1916 I had my first sitting with Mrs. Leonard, who certainly knew nothing about me. Her control 'Feda' (a control is a kind of familiar spirit, probably what is known as a 'secondary personality', of the medium) gave the name H—— (not a very common one) correctly, which is rare at a first sitting; she then described a game which she said he was 'showing' her, '. . . a funny game . . . you play it with a ball, but you don't hit it . . . you *push* it with a stick'. The reference is obviously to billiards, which D. G. H. and I played together a great deal when we were up at Cambridge just before the war. Neither of us played much with anyone else, and the game is not one which most people would guess. Finally, she kept on saying, 'He's showing Feda *roses*, heaps of *roses*—Feda doesn't know what it means, but it's *roses*'. The fact was that, during the aforesaid period of our intimacy at Cambridge, I had been somewhat flagrantly jilted by a girl named *Rose* (pseudonymous, as are the roses above, but a flower name) and had discussed this distressing event exhaustively with D. G. H. Taken collectively, it would have been difficult to improve on these items as evidence of my cousin's identity.

(*a*) A group of experimenters of the American Society for

Psychical Research, using 'table-tilting', obtained the following details from an ostensible communicator: her name was Florence, her father was a Congregational minister named Cyrus Richardson of Nashua, New Hampshire, she had a married sister with a 'jewel' name. All these points were verified, the sister's name being Pearl.

In another case the communicator gave her name as Mrs. Darrow, of Painesville, stated that her husband's name was John, that she had a deceased daughter, and that her old 'Home is now Elks' Home'. These details were also verified, although (as in the previous case) none of the experimenters had ever, so far as they knew, heard of any of the persons concerned.

The evidence is good, though the items were more or less public property, as compared with the more intimate personal details referred to in my own case. On the other hand, for the whole series of experiments, no fewer than 105 points of identification were verified out of 156 given, 35 were unverifiable, and only 16 were erroneous. Thousand-to-one coincidences do, of course, occasionally occur, but only about once in a thousand times on the average, if no factor other than chance is involved.

(*c*) Dr. S. G. Soal, a well-known and exceptionally experienced investigator, obtained through Mrs. Blanche Cooper a message purporting to come from his deceased brother Frank. This stated that he had buried a heavy 'medal', which he used to carry on a chain, near the brick fireplace of a hut built by the brothers when they were boys, some ten or twelve years previously. Excavation on the site revealed, at the indicated position, a disc of *lead* about two inches in diameter and about a quarter of an inch thick; no chain was found, but a hole was bored near the edge of the disc as if for suspension by a chain or string. This again is good evidence, since the information given is most unlikely to have been known to anyone but the supposed communicator.

Apart from such simple and straightforward cases, there are others, too complex even to summarize here, such as the Cross-correspondences, Literary Puzzles, etc., particularly studied by members of the English Society for Psychical Research, which show a high degree of specialized knowledge (notably classical scholarship) and of apparent planning as test cases.

In short, it is not a matter of whether the available evidence is good or not—it is excellent—but of what is logically implied if we accept this evidence.

It is commonly assumed (here is the third error) that a man's 'mind' or 'personality' is a single indivisible unity, so that the question whether it has survived death or not admits of, and indeed requires, an all-or-none, yes-or-no answer. But, if there is one thing that modern psychology does make clear, it is that this assumption is untrue. On the contrary, it displays the mind rather as a loosely federated system of what we may roughly call 'ideas' and 'thoughts'—moods, complexes, sentiments, etc.—forming sub-systems of varying complexity and coherence, some of which in certain cases are quite capable of breaking away from the main mass and, so to say, setting up on their own.

A classical example is that of Miss Beauchamp, normally a middle-aged lady of staid and regular habit, who developed a strongly contrasting 'secondary' personality, known as Sally, lively and mischievous, which took a peculiar delight in doing whatever would most tease and annoy the normal 'self'. The case was actually much more complicated than this, but under treatment the various 'secondaries' were caused to coalesce, and the 'real' personality reconstructed.[1]

Once we have grasped this point and realized that to speak of a Mind or Self or Ego, apart from the idea-systems that compose it, is quite meaningless—because inherently unobservable—it is easy to see that the true question is not so much whether the ideas, memories, etc., composing the federated system continue to exist after death (as the evidence indicates that they do), as whether the system continues to hold together as a whole, or is liable to suffer some greater or less degree of disintegration—as some of the evidence also suggests. It is a question of the stability of the system as an

[1] See Morton Prince, *The Dissociation of a Personality*.

organic structure, rather than of its total annihilation or perfect persistence.

From a certain point of view the suggestion that the characteristic idea-system that we call our personality may gradually disintegrate, is likely to cause alarm and despondency; but I am inclined to think that this involves the greatest error of all—the error of supposing that its preservation is important or praiseworthy or desirable.

We cling to our individuality, or 'I-consciousness', mainly, I think, because it is of elementary biological value to do so. But, if we reflect more carefully, we shall probably realize that this I-consciousness is a handicap and a limitation—not to say a vanity and a snare.[1] The worst of the distresses that afflict us always seem, on analysis, to be a matter of conflict between opposing parts of our 'selves'; and it is in our worst moments, when despair and self-pity have us by the throat, that we are most conscious of these 'selves', as beings cut off from human kind and victimized by a cruel conspiracy of Fate and of our fellow men. But when we are intensely concentrated on any activity, or in any contemplation, the sense of I-ness fades out—eminently we remain *conscious*, but not *self-conscious*—and it seems to me that there is much to be said for the view that this un-self-troubled state is just about the ideal one to be in;[2] and I strongly suspect that our clinging to our I-ness is not much more than the sentimentality and laziness that makes us cling to a disreputable pair of old and none too comfortable boots to the pinching of which we happen to have grown accustomed.

At any rate it seems clear to me that it is no use spending this life dwelling in hope or fear upon the prospects of the next, or in wondering whether we shall go on being the same person then as we are now; or even in agonizing over whether we shall or shall not meet So-and-so again as we knew him

[1] This paragraph is irrelevant, I think, to the objection stated in the previous one. The disappearance of *self-consciousness* may be highly desirable (at any rate in some sense of the term 'self-consciousness'). It does not follow that the disintegration of the mind itself—i.e. of the system of ideas, memories, etc.—is therefore desirable. (H. H. P.)

[2] Cf. Ch. V, para. 72, above, pp. 181–41. (H. H. P.)

here. For purely technical reasons, I think the answer to the last question is that we almost certainly shall—for so long and in such degree at least as it remains important to us to do so. But even here there is much force in Walter Leaf's contention, written so long ago as 1903, when he says 'to me "personality" presents itself mainly . . . as the barrier which inexorably cuts me off from those who are nearest and dearest to me, so that they can never know half the reasons why I smile or sigh.'[1]

The lesson to be drawn from what we so far know of survival is, I think, not that we should eat and drink for to-morrow we die, but that we should seek to free ourselves (so far as the politicians, the economists, and the atom-bombers will let us) from that tangle of vanities and fears that make up our so treasured and self-adulated 'selves'.

[1] *Proceedings* of the Society for Psychical Research, Vol. XIII (1903–4), p. 61, in a review of Myers' *Human Personality*.

III

DOES TO-MORROW EXIST?

'What a ridiculous question! If to-morrow existed it would be to-day, wouldn't it? Of course it can't exist . . . what on earth are you talking about?'

But although the reader's reaction to my title is natural enough, the question is by no means so silly as it sounds; indeed, it is one which we shall be obliged to answer sooner or later, if we are ever to understand the nature of the human mind and its place in the universe—and until we do that, we shall presumably go on making as big a muddle of our lives as we have done up to date; and this we would, I am sure, all wish to avoid.

The problem, which has a certain fascination of its own, apart from this long-term view, arises from the fact that recent work has placed the phenomena of 'precognition'—that is to say, roughly, the 'foreseeing' of future events—on a firm experimental basis, so that it can no longer be brushed aside as all due to coincidence, misremembering, and so forth. But, since you obviously cannot 'cognize', or be aware of, or indeed have any sort of relation to, anything at all unless it exists, it is clear that the so-called 'future' event precognized must, in some sense at least, exist at the moment of precognition. A terrible dilemma, to which I myself see only one possible answer—and that is not to be found by blethering (if I may be allowed the term) about the Unreality of Time (which means nothing at all) or The Eternal Now—which does not, I fancy, mean anything more.

From the earliest times men have longed (perhaps imprudently) to know the future, and a long line of prophets, seers, oracles, and augurs have catered, more or less honestly, for the demand. On the whole, despite a good deal of inevitable charlatanry, I should say that the history of the subject has been relatively respectable and compares favourably with that of any other 'paranormal' activity, such as necromancy, clairvoyance at a distance, and so forth. From classical times onward there has been reported a number of instances of prophecies strikingly fulfilled, and in recent years numerous smaller but not less cogent cases have been collected by the Societies for Psychical Research and by private students.[1] To give but a single instance, which has at least the merit of being picturesque: Mrs. Atlay, wife of a former Bishop of Hereford, dreams that she reads family prayers in the hall, instead of in the chapel, as her husband is away, and that after doing so she finds in the dining-room a large pig standing between the table and the sideboard. She comes down and relates her dream *before* reading prayers. It is precisely fulfilled, including the position of the pig. Note that the pig was safely in its sty at the time of dreaming, but escaped while prayers were being read. Pigs are, one may suppose, so seldom found in episcopal dining-rooms that it is difficult to attribute such an occurrence (which is well authenticated) to mere coincidence; and the difficulty becomes a virtual impossibility when such cases are multiplied by the score.

On a different line, Mr. J. W. Dunne created a considerable stir with his book *An Experiment with Time* (first published 1927) describing how he found that many elements in his own dreams were apparently precognitive in character.[2]

But such more or less spontaneous cases, though extremely impressive, are very difficult to assess, and therefore insufficient by themselves to overcome the strong psychological resistance that the idea of precognition not unnaturally encounters in most minds.

More recently, various investigators, notably Rhine, Soal and Goldney, Tyrrell, and the present writer,[3] have obtained

[1] See, for example, H. F. Saltmarsh, *Foreknowledge* (Bell, London, 1938).

[2] J. W. Dunne, *An Experiment with Time* (Black, London, second edition, 1929). These results were supplemented by a formidable theory involving successive 'dimensions' of Time, in an infinite regression; but I do not think that this has been accepted by anyone competent to analyse it.

[3] J. B. Rhine, *The Reach of the Mind* (Faber & Faber, London, 1948), Ch. V; S. G. Soal and K. M. Goldney, *Proceedings* of the S.P.R., Vol. XLVII, Pt. 167, December, 1943. G. N. M. Tyrrell, *Science and Psychical*

significant (i.e. non-chance) precognitive results under strict experimental conditions. The classical work is that of Soal and Goldney, in which, briefly, their subject was asked to 'guess' cards drawn at random, under elaborately precautionary conditions, by the experimenters; they found that he scored significantly not on the card known to the experimenter at the moment of guessing, but on the *next* card in the series, i.e. one the nature of which was not known to anyone at that moment. The work was far too careful for the results to be attributed to faulty procedure, and is well supported by other researches; and there can be no doubt that, taking the evidence as a whole, we must accept it as a fact in nature that a future event, such as the observation of a randomly selected card, can be, in some manner, 'foreseen' or 'precognized', at least by certain people under certain conditions.

But how? In particular, how is it possible for a physical event (i.e. some configuration, so to say, of material objects) to 'exist' in such a sense, or to such an extent, that it can be cognized, and yet *not* to 'exist' in the sense that, as we say, it has not happened yet? No satisfactory explanation has yet been given, and it will at least be good, clean fun, for those who like this kind of thing, to see whether we can do better than heretofore.

Before we can understand what is implied by 'foreseeing', 'precognizing', or 'preperceiving' (if I may coin a term) the form that a future event will take, or the characteristics an object will display, we must examine what actually happens when we 'see', or 'cognize' or 'perceive' an event or object in the ordinary way. If I say 'I see a tomato', I am using a highly condensed form of words, which is taken to imply (and usually correctly) much more than the facts actually warrant. More strictly, I ought to say something like 'I am aware of a red patch, of such-and-such a shape, hue, and shading; and experience of similar red patches in the past leads me to expect that, if I stretch out my hand and grasp it, I shall be aware of certain feelings of smoothness, coolness, firmness, etc., that if

Phenomena (Methuen, London, 1938), Ch. VIII; W. Whately Carington, *Telepathy* (Methuen, London, 1945), Ch. III.

I squeeze hard I shall be aware of certain sensations of yielding, moisture, etc. ('squashing'), that if I put in my mouth and bite it I shall be aware of . . . etc., etc.' Even this is not a complete expansion, but it will serve our present purpose.[1]

Note, all-importantly, that if the thing does not behave in this way—to put it very colloquially—that is to say, if this sequence of awareness, etc., is not followed with reasonable exactitude, it is not a *tomato* that I am seeing; it may be a wax imitation, or maybe I am enjoying an hallucination. A 'real' tomato, as we put it, is that which conforms to the accepted specification of a tomato.[2]

But communication would be quite impracticable if we had to use this enormously expanded form of words every time we wished to speak of a material object, and we accordingly use the shorthand symbol or 'portmanteau' phrase 'a tomato' for all ordinary purposes; but it will be misleading unless it has all the above expansion packed into it, however little we may realize this when talking.

Note next, as of equal importance, that all we directly know at first hand are the sensations of red, smooth, cool, firm, moist, salty, acid, etc., and, of course, those of stretching, grasping, biting, etc. The first lot (together, I need hardly say, with those corresponding to any other operations we may perform) taken collectively *are* the tomato; the tomato *is* this collection of sensations, or awarenesses, etc.

But, you may object, these are only properties, etc., 'of' the tomato—what about the tomato *itself*? This is the vulgar error into which pretty well everyone has fallen since the beginning of time, namely, of supposing that because we find it convenient to use the phrase 'the tomato' as a time-saving shorthand symbol, there must be a *thing* (tomato itself) for which the word stands; and that this 'has' the properties, or 'causes' the experiences, etc., which are held together in their observed pattern, so to say, by their relationship to *it*—almost as labels might be stuck on a suitcase.[3] No conceivable process

[1] Cf. Ch. IV, Sects. 38 and 39, pp. 94–7, above. (H. H. P.)
[2] Cf. Ch. IV, Sect. 51, p. 125, above. (H. H. P.)
[3] Cf. Ch. IV, Sects. 44 and 45, pp. 104–9, above. (H. H. P.)

of observation, however, will enable you to discover this alleged 'tomato-itself'; all you can do is to discover new tomato-properties; and to talk about anything which is inherently unobservable is to talk meaningless non-sense.

The moment we abandon this primitive superstition about 'things-themselves' we are free to realize (given a little resolution and fortitude) that there is no compulsive necessity which requires that visual, tactile, gustatory, etc., experiences should be related in the particular way known as constituting a material object. There is no coercive reason why the various groups should not function and be cognized independently; and, as a matter of observed fact, they sometimes do and are. When we cognize a visual group, say, not related in the material-object way with the expected tactile group, we say that we are having a visual hallucination; and such experiences are, as is well known, not very uncommon, while tactile, auditory, and other forms of hallucination, though rarer, are by no means unknown. We say that there is an event in the physical world, or that a material object is present, only when the various groups are (I am tempted to say 'happen to be') present together, or coincident, or some such phrase, much as we only have a coloured print when the three components of the three-colour process are superimposed.

Now we can clear up the basic difficulty about precognition. I have, say, a vision—sufficiently detailed to exclude coincidence—of some future event; that is to say, an event later occurs of which the visual appearances closely resemble my previous experience. The visual components of that event clearly did exist at the time I had the vision, and that covers the difficulty about the impossibility of cognizing something that does not exist. But they had not been then joined by—if I may put it so—or become coincident with the tactile, auditory, etc., components in the relational pattern that constitutes the occurrence of the event, and this deals with the point about an event not existing before it occurs.

It all seems to me perfectly simple and straightforward, provided we are not scared of sticking to the facts, and refuse to be led astray by a lot of verbal balderdash about 'things' which 'have' properties, and similar pitfalls arising from the uncritical use of language.

I think the view suggested also gets us out of the difficulties arising from the view that, if the future can be foreseen, it must be fixed and immutable, so that we cannot avoid it, but are deterministically predestined to endure whatever this now-existing future holds for us. Or, as Dr .Joad plaintively remarked, 'If the future exists, what is to become of Free Will?'

Personally, I have never been able to understand what the term 'Free Will' is supposed to mean (and very much doubt whether anyone else does).[1] Determinism I know, and Chance I know—well, more or less—but Free Will seems to me to be wholly meaningless; for whatever deterministic compulsions of external origin may be removed, you must surely still act or decide either purely by chance, or else for reasons of one kind or another, and these reasons will be just as determinative as anything else. Actually, I think it is just another of these word-generated false problems, which are merely nonsensical and mean nothing at all; because I cannot conceive of any possible means of distinguishing a Deterministic world from a Free Will world, and, if you cannot do that, your verbal distinction is just so much empty noise.

But, in any event, it seems quite clear that if the factual occurrence of an event consists in the coming of certain components into a certain relation, of coincidence or the like, then the fact that one of these components has been perceived or cognized does not in the least imply, as a logical necessity, that the coincidence will take place and the event occur.[2] So we need have no qualms, so far as I can see, even on the

[1] Cf. Ch. V, Sects. 75–6, pp. 189–98, above. (H.H.P.)

[2] i.e. that tactual and other cognita will be *combined* with these visual ones, to make up a complete physical event. Thus an unfulfilled precognition would be exactly analogous to an ordinary hallucination—except that the hallucinatory sense-datum would be future instead of present. In a fulfilled precognition, however, this same sense-datum ought to be *re*-experienced when the appropriate moment of time arrives. Does this in fact happen? Or are we to suppose that at the appropriate time there will be a visual 'cognizable' of the appropriate sort, even though no one happens to sense it? (H. H. P.)

non-logical ground of our distaste for determinism, to accepting precognition as a fact. And, if we do so and follow the matter up by suitable experiment and reasoning, we shall, I believe, be well on the way to shaking what Professor Lindemann (now Lord Cherwell) once called 'the grim pre-eminence accorded by age-long consent to Time'. And when we have done that, we may begin to feel a little more at home in the universe than we do at present.

INDEX